T0365959

A Nobel Prize winner, Professor Tinbergen has spent a lifetime exploring the behavior of many types of animals in their natural environments. His work has been characterized as a "breath of fresh air" in fields that were in danger of losing touch with nature and, in following Tinbergen on what he calls his "aimless wanderings," the reader will catch a unique glimpse into the workshop of ethology. Even when reporting on sophisticated experiments or when developing new theoretical concepts and arguments, Tinbergen writes simply, lucidly, and precisely.

The present volume spans forty years of pioneer investigation and includes selections on the behavior of gulls; on the homing, landmark preference, and prey findings of the digger wasp; on the food hoarding of foxes; and on creatures' living scattered as a defense against predators. These are classic original studies which will fascinate the increasing number of readers interested in the topical problems of animal and human behavior.

Frontispiece: A calm early morning well suited to track reading (see chapter on 'Food Hoarding by the Fox'). The author is on the right. (***Photo****: Lawrence C. Shaffer, Dept of Zoology, Oxford.*)

The Animal in its World

NIKO TINBERGEN FRS

Professor of Animal Behaviour and Fellow of Wolfson College, Oxford

The Animal in its World

Explorations of an Ethologist

1932–1972

FOREWORD BY SIR PETER MEDAWAR, FRS

Volume One

FIELD STUDIES

HARVARD UNIVERSITY PRESS

CAMBRIDGE, MASSACHUSETTS

© George Allen & Unwin Ltd, 1972

ISBN 0-674-03725-1 (cloth)

ISBN 0-674-03724-3 (paper)

Library of Congress Catalog Card Number 72-94876

Third Printing 1975

PRINTED IN THE UNITED STATES OF AMERICA

Dedicated to E.A.T.
for her interest, encouragement
and tolerance

Foreword

BY SIR PETER MEDAWAR, FRS

Niko Tinbergen is one of the grand masters of Ethology, and the papers published here are among its most important documents: they are a source-book for students of animal behaviour and will give the historian of Ideas an insight into the early days of one of the most influential movements in modern science.

Anybody who thinks that Ethology consists of a passive imbibition of the information proffered by nature still has much to learn. The first stage in a behavioural analysis is, of course, to observe and record what is actually going on. This will involve intent and prolonged observation until what an untrained observer might dismiss as a sequence of unrelated behavioural performances is seen to fall into well-defined and functionally connected sequences or behaviour structures. These behaviour structures do not declare themselves in any obvious way. Their identification depends upon an imaginative conjecture on the part of the observer which further observation may or may not uphold. As in other branches of science, this is a creative process in which the imagination must take the initiative. Another important element in the ethological approach is the comparison of behaviour structures among different but related animals, which in turn opens out the possibility of identifying homologies of behaviour. The word 'homology' is not easy to define in any context, but in an ethological context it may be exemplified by saying that the behaviour associated with mother-love and suckling is obviously homologous or genetically cognate in man and in apes. Clearly this complex behavioural repertoire did not spring into being fully formed with the inception of the species *homo sapiens*.

In Tinbergen's early days as a research worker 'experimental' was the boss word, much as 'molecular' is today. Everything had to be experimental: Embryology, Pathology, Physiology, and, if possible, the study of behaviour. Today anybody who studies 'molecular' something instead of humdrum old everyday something feels an unaccountable increase of stature. Some of the early critics of Ethology believed that, because of its mainly observational character, it could not possibly be really in the mainstream of biological research. The idea that there is something *essentially* meritorious about experimentation has been carried over from Bacon's original usage of the term[1] and the advocacy that went with it. It just so happens, however, that some of Tinbergen's earliest work was experimental.[2] Much of the success of his work is due to his adoption of a judicious blend of observation and critical experimentation.

Ethology was soon recognised as one of the really important developments in modern biology, and became the subject of excited discussion in zoological departments. Many of the brightest zoological students and natural historians cherished ambitions to study under Tinbergen in Oxford. I say 'in Oxford', although Tinbergen was Professor of Zoology in the University of Leiden, because one of Professor A. C. Hardy's many bright ideas had been to persuade Tinbergen to come to work in Oxford and enjoy the facilities and spacious academic atmosphere he felt sure Oxford could provide. By juggling with half-vacancies, as we shall soon all be juggling again, a post was created which Tinbergen agreed to take. (Harvard wooed him too, but Tinbergen has not regretted his choice.) One thing Oxford could not supply: a seashore; but a Landrover overcame the difficulty and Oxford's rough equidistance from the sea in all directions came to be seen as an advantage by making Oxford literally the centre of Herring Gull field stations.

It is a source of great pride to me that, when funds for Ethology were running low, I was able to interest the Nuffield Foundation in Tinbergen's work which they supported by grants totalling several thousand pounds over a period of ten years, whereupon the support of his research devolved upon the Natural Environment Research Council. I was relieved that the Foundation took my advice instead of that of the eminent neurophysiologist whom they also consulted and who said of Ethology, 'Why, that's just birdwatching, isn't it?'

The paper on Autism (to be included in a later volume) is to be read as a study in the application of ethological methods to human infants. Notice here, for example, the identification of certain elements of autistic behaviour in the behavioural repertoire of normal children, and the implied hypothesis that Autism may be a consequence of overstimulation—of too much intrusive attention perhaps, and not, as some immature psychologists have been inclined to feel, of too little attention. The picture of Tinbergen that comes to mind

[1] Medawar, P. B., *The Art of the Soluble*, Methuen & Co., London, 1967; and also *Induction and Intuition in Scientific Thought*, Philadelphia and London, 1969.
[2] Tinbergen, N., 'Curious Naturalists', *Country Life*, Feltham, Middlesex, 1966; and also *The Study of Instinct*, Oxford, 1951, p. 32.

through this paper, with his enthusiasm, kindness, lack of pretension and acute observation, is entirely authentic.

Tinbergen's own childhood was spent amidst a happy and cheerful family and he showed an early predilection for a naturalist's pursuits. One of his school reports, with an almost uncanny lack of prescience, said that his powers of application were not such as to equip him for a career as a natural historian. Tinbergen spent much of the war in a hostage camp and was able to reflect, not for the first time, that wild animals are less to be feared than malevolent human beings.

Tinbergen himself has rarely attempted to establish significant homologies of behaviour between human beings and lower animals, a subject on which his pupil Desmond Morris has written so skilfully and so divertingly (*The Naked Ape*, London, 1967). 'What is there', we may well ask, 'in all this naked ape business?' The approach can be an eye-opener to those who had not realised that the human behavioural repertoire is of some evolutionary depth. Human behaviour did not spring into being fully fashioned but, like human bodily structures, must have evolved from lowlier precursors. Of human bodily structures Darwin said that, in spite of what he was kind enough to call our exalted powers, 'Man still bears in his bodily frame the indelible stamp of his lowly origin.' Without doubt the same is true of many human behavioural performances. As with many of the so-called 'behavioural' sciences, the difficulty arises at a demonstrative or evidential level. If any ill-disposed critic says of some theorem 'I simply don't believe it', its champion has no recourse except to appeal to the general plausibility and reasonableness of his argument. There is no well-understood and authenticated remedy for disbelief, as there is in the conventional or 'hard' sciences.

Tinbergen has had a number of bright pupils of whom he is very proud and with whom he remains on good terms although Ethology is a controversial and rather passionate subject. Through them and through his own work he achieved the ambition he first formulated in the wartime hostage camp: to bring Ethology to the English-speaking world (in effect to England, because America already had a long tradition of animal behavioural research, although in quite a different style—more 'scientific', in what I think of as the wrongheaded usage of that term: more measuring instruments and many more measurements but less to do with natural behaviour or indeed with nature at all. The terminology that went with it—tropisms, taxes and the like—now have to me as olde-worlde a flavour as any Tea Shoppe in a small English cathedral town.)

Tinbergen sometimes describes himself as an artist *manqué* rather than as a scientist (though the two descriptions are not incompatible), but he is in fact a first-rate scientist whose austere standards of research have steered Ethology safely between the beguilements of imaginative story-telling and the barrenness of compiling inventories of facts.

Contents

Translated from the German originals

Illustrations

Introduction

This book, published at the promptings of several of my friends whose judgment I decided to accept, contains reprints of some of the papers which, partly together with pupils, I have published between 1932 and 1971, when I was finding my feet in the growing science of Ethology. With those friends I believe that this collection can assist in meeting the increasing demand for 'readers' for schools, colleges, and universities, whose students are not satisfied by digests only, but want to have access to some original publications. What made me overcome my diffidence in republishing some of my old work was not only the fact that many of these papers appeared either in German (a language still largely wished away by my English-speaking colleagues) or in relatively inaccessible places, but also that, however aimless my wanderings in these studies seemed to be at the time, I have come to feel more and more that there has been a consistent line in them. I also believe that this line is of not merely historical interest, but is still worth following, and indeed will never become outdated; and that it is worth trying to convince budding biologists that it is a fruitful approach to the problems of the life sciences and of the behavioural sciences in particular, including that of Man's behaviour.

Two aspects of this approach seem to me worth singling out. First of all, I believe that we should continue to observe and describe before we experiment; and second, that with regard to behaviour (as with all life processes) we should ask the question 'what's the use of what the animal does?' 'does it contribute to the animal's success, and if so, how?' as well as the question 'what makes it happen?'

With respect to observation I can say no more than that I have found that in our young science it pays to become acquainted with, and to wonder about, the variety of phenomena to be seen before one turns either to experiment, or to generalisation. To some this 'watching and wondering' comes naturally; others find it more difficult to curb their impatience. The experimenter and the generaliser tend to consider mere observation and description 'unscientific'; the observer tends to find the other man narrow and impatient. Like many other sciences, those concerned with behaviour have swung from phases of predominantly observational and interpretative activity to periods of penetration into specific phenomena, considered representative of general problems. I believe that Ethology is at the moment swinging away from observation, and that it is time for another move, a return towards a better balance between the two tendencies. Looking at this collection of little probes into a variety of phenomena and problems, I have the temerity to believe that they demonstrate how fruitful the 'seesawing' between observation and experiment can be; many readers will recognise in several of my papers the germs of much painstaking modern research by others.

Naturally, observation is always much more than a passive taking in of the outside events; as many authors have pointed out, all observation is selective; and this selectiveness is determined from within. Popper has repeatedly stressed this; he dislikes the word 'inductive', and stresses the *naiveté* (and the futility) of believing that the observer is a *tabula rasa*. I believe that observers and experimenters can find common ground in acknowledging that we all start with expectations, with prejudices, even with hypotheses, and that we begin to wonder when we find, either at the observational or the experimental level, that what we observe is contrary to our expectation; we are amazed at the refutation of what we expected, and it is this amazement that spurs us on.

This brings me to my second point: what is it that guides the biologist, and therefore the ethologist when he observes? I believe that it is worthwhile trying to describe this in the simplest possible terms: biologists are drawn to study events that *seem* to contradict what we have been taught to expect on the basis of our knowledge of non-living things. It is this discrepancy between what an animal 'ought to do' and what it is actually seen to do that makes us wonder. Like a stone released in mid-air, a bird ought to fall; yet it flies away. An animal resists, evades, utilises its environment instead of meekly submitting to its influences; rather than move towards increasing stability and probability, it survives, reproduces, grows and evolves. Yet we do not *believe* that it does all this against the 'laws' of Physics—our *faith* is that upon closer investigation it will turn out to be a special case. This contradiction between what Physics makes us expect and what we actually see gives Biology an extra dimension; it leads to the question 'what is this life process *for*?'; 'what is its use?' Provided we apply the word 'use' in a limited sense—that of 'ensuring survival'—this question is legitimate in Biology, whereas in Physics it is out of bounds—it is 'metaphysical' to ask 'what's the use of the planetary system, of an atom, of the tidal movements?'

This validity of the question of the 'survival value' of what an animal does is the second point that this collection of reprints is intended to illustrate. In all the papers a preoccupation with the animal's struggle for success is evident. From the start, the observations lead to two questions rather than, as in the physical sciences, to one. One finds oneself turning all the time from asking 'what's its use?' to 'how is it done?', and vice versa. One proceeds either by studying the effects of observed behaviour, and finding out how these effects influence success; or one spots a potentially destructive or obstructive environmental pressure, and tries to find out how this pressure is met.

This dual procedure, of which I have myself only gradually become conscious, leads to purposefulness and economy of research effort. The study of environmental pressures leads one to single out the truly biological problems and saves one from studying mechanisms for their own sake—as it were *in vacuo*; and the question 'what's its use?' leads one to explore the pressures that the environment places in the animal's way. My reprints are intended as a plea for continuing to investigate the link between behavioural mechanisms as found, and their relevance for success.

What emerges is in essence an increasingly detailed picture, at the behavioural level, of the close, or perhaps rather the sufficiently close correlation between what is required for success and what is done; and so to an increasing insight into, and respect for, what natural selection has obviously produced. That natural selection can and does produce evolutionary changes cannot of course be denied—in the search for what it has actually produced the best one can do is to try to spot seeming contradictions and find out whether they really *are* contradictions, or turn out upon analysis to be consistent. How illuminating this is must be pondered over by the reader; as will be discussed in papers 16, 17 and 18, I myself believe that its value will become clear when we begin to apply this type of work to man and his unique feature: the loss of adaptedness as the result of his cultural evolution. In putting our own, frighteningly ravaged house in order, a sound judgment of adaptedness as the result of genetic evolution, and loss of adaptedness in a culturally changed environment, seems to me crucial. For this, an understanding of our niche and of our original adaptedness, in as much concrete detail as possible, will be indispensable—general statements will not be sufficient.

Whether we shall ever find the solution to our dilemma, and be able to show that at *all* levels of integration, from animal communities down to molecules, there is no contradiction between the laws of physics and life processes no one can tell, although such a fusion between Physics and Biology seems to be within sight at the molecular level. But what does seem to emerge is something of beauty—it gives immense satisfaction when even the most improbable things an animal does can be seen to meet what the natural environment demands; the environment which has given and is giving direction to evolution.

It is a pleasure to thank Miss L. Gombrich for translating papers 4, 9, 14, and 16; Dr Anne Rasa for translating papers 2, 3, and 5, and Dr R. D. Martin for translating paper 11; Mrs P. Searle for secretarial assistance of many kinds; Mrs (Dr) M. Dawkins for a great deal of help ranging from critically reading and polishing the entire text to much tedious editorial work; Miss C. Court for redrawing a number of illustrations and finally Mr Charles Furth, of Allen and Unwin, for his encouragement and patience.

That my friend Peter Medawar, to whose support not only I myself, but all British ethologists owe such a great deal, has written a foreword to this volume, means more to me than I can say.

References to the Original Papers

SECTION I

1 N. TINBERGEN (1959). 'Comparative studies of the behaviour of gulls (Laridae): a progress report', *Behaviour*, **15,** 1–70.

SECTION II

2 N. TINBERGEN (1932). 'Ueber die Orientierung des Bienenwolfes (*Philanthus triangulum* Fabr.)', *Zs. vergl. Physiol.*, **16,** 305–34.

3 N. TINBERGEN (1935). 'Ueber die Orientierung des Bienenwolfes (*Philanthus triangulum* Fabr.) II. Die Bienenjagd', *Zs. vergl. Physiol.*, **21,** 699–716.

4 N. TINBERGEN und W. KRUYT (1938). 'Ueber die Orientierung des Bienenwolfes (*Philanthus triangulum* Fabr.) III. Die Bevorzugung bestimmter Wegmarken', *Zs. vergl. Physiol.*, **25,** 292–334.

5 N. TINBERGEN, B. J. D. MEEUSE, L. K. BOEREMA und W. VAROSSIEAU (1942). 'Die Balz des Samtfalters, *Eumenis semele* (L.)', *Zs. Tierpsychol.*, **5,** 182–226.

6 N. TINBERGEN, G. J. BROEKHUYSEN, F. FEEKES, J. C. W. HOUGHTON, H. KRUUK and E. SZULC (1963). 'Egg shell removal by the Black-headed Gull, *Larus ridibundus* L.; a behaviour component of camouflage', *Behaviour*, **19,** 74–117.

7 N. TINBERGEN, H. KRUUK and M. PAILLETTE (1962). 'Egg shell removal by the Black-headed Gull (*Larus r. ridibundus* L.) II. The effects of experience on the response to colour', *Bird Study*, **9,** 123–31.

8 N. TINBERGEN, H. KRUUK, M. PAILLETTE and R. STAMM (1962). 'How do Black-headed Gulls distinguish between eggs and egg-shells?', *British Birds*, **55,** 120–9.

9 N. TINBERGEN (1965). 'Von den Vorratskammern des Rotfuchses (*Vulpes vulpes* L.)', *Zs. Tierpsychol.*, **22,** 119–49.

10 N. TINBERGEN, M. IMPEKOVEN and D. FRANCK (1967). 'An experiment on spacing-out as a defence against predation', *Behaviour*, **28,** 307–21.

Section I
Observational and Interpretative Studies

Author's Notes

My original plan was to open this section with one of a number of single-species studies which I published at the time when work on 'ethograms' formed the backbone of Ethology. Following in the footsteps of such men as Heinroth, Huxley, Verwey, and, of course, Konrad Lorenz, I browsed, so to speak, in a variety of habitats, often together with pupils and colleagues, and a number of publications appeared in the 'thirties, ranging from short articles mainly on various birds, fish, and insects (Sticklebacks, Newts, Bitterlings, Water Beetles, Owls, Gulls, Terns) in Dutch journals, to more or less monographic studies published in international periodicals: papers on the Hobby, the Red-necked Phalarope, the Snow Bunting, and finally a book on the Herring Gull.

When it came to making up my mind which of these to reprint here, I decided reluctantly that none of them really qualified any more. All of them are too lengthy, and contain a mixture of rather anecdotal information and unsophisticated, exploratory excursions into problems on which much better work has been done since; moreover, the gist of most of these studies has already been summarised in my previous books. Yet they are worth mentioning because, as a teacher, I feel that I ought to pass on to younger readers my experience that to each budding ethologist such periods of exploratory, intuitive observation are of immense value. They remain so throughout one's research career—in particular when one feels in danger of getting out of touch with the natural phenomena; of narrowing one's field of vision.

It is partly because of the exploratory nature of these earlier papers, partly because of their bulk, that this first section contains only one paper, published in 1959. This paper, though factually of course not up to date, was selected not only because it is representative of largely observational studies but also because it has a dimension hardly found in my earlier papers: inter-species comparison.

More precisely, what was lacking in all those one-species probes was any serious attempt to apply the comparative method for the purpose of interpreting diversity and similarity as the outcome of adaptive divergence and convergence; as such they show a failure to take in the lessons I could have learned from Whitman, Heinroth, Huxley, Verwey and Lorenz.

It was Konrad Lorenz's relentless prodding which made me turn to truly comparative studies. Having studied the Herring Gull in some depth, I naturally turned (finally convinced by Lorenz's paper on dabbling ducks) to other gulls. Being interested in animals in their normal environment, I had to go where the various species lived. Such an ambitious programme could not be carried out by one person, and much factual material on which the summarising paper draws was collected by pupils. My interest in the function of behaviour steered these studies rather in the direction of understanding adaptive radiation; and at the start our attention was concentrated on the most puzzling behaviour patterns, the 'displays'. But the paper shows very clearly the influence of team work of the kind where the leader is not concerned to steer his pupils' work too closely; I have in that time learnt at least as much from my co-workers as they from me; of this the classical study of Dr Esther Cullen on the Kittiwake is perhaps the most striking example.

My 'progress report' does not bring out sufficiently the debt owed by this project to Comparative Anatomy; had it done so it would once more have become clear that much of what Ethology has done is no more than the application of already practised biological procedures to behaviour rather than to structure. But the 'rationale', the spelling-out of some of our guiding lines, will reveal to those at home in Comparative Anatomy that we were not doing anything really new.

This consideration, that what Ethology has been doing is in essence not much more than applying already accepted methods to the set of life processes called behaviour, suggests at the same time another justification for republishing old papers; I am convinced that our work will in turn be considered as fruitful to students of human behaviour and human sociology, fields in which the application of biological ways of thinking is (understandably in view of the complexity of the phenomena) a long way behind Ethology.

(From the Department of Zoology, University of Oxford)

1

Comparative Studies of the Behaviour of Gulls (Laridae)

A Progress Report (1959[1])

I Introduction

Stimulated by Heinroth's (27) and Lorenz's (45) studies of the behaviour of Anatidae, members of the Research Unit of Animal Behaviour of this Department started, in 1950, a programme of comparative studies of the behaviour of gulls. Foundations had already been laid by the work of Strong (71), Portielje (67), Goethe (24) and Tinbergen (76, 79) on the Herring Gull, and by Kirkman (38) on the Black-headed Gull. The aims of our studies were: first, a description of the behaviour of as many species as possible; as complete a coverage as possible of the entire behaviour pattern of each species; and third, analyses of the functions, the causation and the origin of the displays, with the ultimate aim of understanding how they could have originated and diverged in the course of speciation. This wide scope of phenomena and of problems was judged necessary because it had become increasingly clear that an understanding of the origin and evolution of displays can be considerably enhanced by a wide knowledge of entire behaviour patterns and by some insight in the functions and the causation of displays.

The present paper attempts to give a summary of the results so far obtained, with a discussion of some problems of function, of motivation, of some evolutionary aspects, and of methods of study. It is based mainly on the following species:

1. The Herring Gull, *Larus argentatus* (the authors mentioned above; Moynihan, **60a**) and its close relatives the Lesser Black-backed Gull *L. fuscus* (Paludan, **63**; J. M. and E. Cullen, unpubl. observations), the Glaucous-winged Gull, *L. glaucescens* (Veitch and Booth, **88**; unpubl. observations of my own), the

[1] For full references to the papers reprinted here see p. 22.

Western Gull, *L. occidentalis* (unpubl. observations of my own).
2. The Common Gull, *L. canus* (Weidmann, **90**) and its relative the Ring-billed Gull, *L. delawarensis* (Moynihan, **60 a, b, c**).
3. The Black-headed Gull, *L.* (subg. *Hydrocoloeus*) *ridibundus* (Moynihan, **58**; Tinbergen and Moynihan, **86**) and other hooded gulls: Bonaparte's Gull, *L. philadelphia* (Moynihan, **60 a, b, c**), Franklin's Gull, *L. pipixcan* (Moynihan, **60 a, b, c**), Laughing Gull, *L. atricilla* (Moynihan, **60 a, b, c**), Hartlaub's Gull, *L. novae-hollandiae hartlaubi* (Tinbergen and Broekhuysen, **85**), Little Gull, *L. minutus* (Moynihan, **58**).
4. The Kittiwake, *Rissa tridactyla* (E. Cullen, **18** and unpubl. observations; Paludan, **64**).
5. The Ivory Gull, *Pagophila eburnea* (Bateson and Plowright, unpubl.).

Almost all studies were conducted in the field; some observations made in zoological gardens have been added.

Films were made of the displays of the Herring Gull, the Lesser Black-backed Gull (N.T.), the Glaucous-winged Gull (N. Hancock, of Broadview, Ill.), the Western Gull (Dr R. Boolootian of Los Angeles), the Black-headed Gull (N.T.), Hartlaub's Gull (Dr G. J. Broekhuysen of Cape Town), the Kittiwake (N.T.) and the Ivory Gull (C. Plowright of Cambridge). Copies of all these films, as well as of a film of the Great Skua, *Catharacta skua* (G. Paris of Gouda, Holland) are in the Department of Zoology and Comparative Anatomy, Oxford. We have found these films of increasing value as the work progressed.

The illustrations in this paper were for the greater part drawn from these films and from still photographs.

Since the work thus deals with many species, and each species has a highly complicated behaviour repertoire; and because objective recording of movements is a much more laborious task than that of structures, even the initial descriptive phase of the work is an extremely time-consuming task, which can by no means be considered to have been completed even for those species we studied most intensely. Further, work on functions and causation is hampered by serious difficulties; and lastly, evidence on evolutionary problems is naturally indirect. For these reasons our work is still far from complete. Yet this progress report seems justified for two reasons. First, several conclusions have been reached which, while tentative, seem to me to be of some general interest; for instance, I believe that we have made definite progress with the analysis of the functions of displays, of some aspects of their causation, and of their evolutionary history;

and I want to call attention to some ideas which might be checked in other animals. Secondly, I am convinced that this kind of work is still hampered by a certain lack of precision in our formulation of problems, of methods, and of conclusions. However much Lorenz (**43**, **45**), Hinde (**32**), Baerends (**5**), Andrew (**1**, **2**), Marler (**48**, **49**), Moynihan (**59**), Morris (**55**, **56**) and others have contributed to theoretical clarification, and however much of great interest has been found already, we are still far from knowing exactly what we are doing; much of our work is guided by intuition rather than by conscious and systematic thinking. This is not surprising since this work is not only concerned with the causation of behaviour, where it is notoriously difficult to apply strictly biological methods of study, but since, in addition, the problems of causation, function and evolutionary history are so closely interrelated that the danger of circular arguments is great. Some space has therefore been given to attempts at reformulation of problems, methods and arguments with a view of further preparing the ground for future work.

I have tried to write this paper for the specialists as well as for the interested zoologists, ornithologists and psychologists. This gives the paper certain characters of a compromise.

I want to express my gratitude to the Cumberland County Council, the Farne Islands Committee, the National Trust, the Nature Conservancy, Sir William Pennington Ramsden, and the Earl of Verulam for permission to work in various gull colonies protected and managed by them; to the Ford Foundation, the National Science Foundation, the Nature Conservancy and the Nuffield Foundation for financial support: to Professor Sir Alister Hardy FRS for encouragement and support of various kinds; to P. Bateson, Dr E. Cullen, Dr J. M. Cullen, G. Manley, Dr M. Moynihan, Dr A. C. Perdeck, C. Plowright, Dr R. Weidmann and Dr U. Weidmann for permission to use unpublished data; and to Dr A. Cain, Dr E. Cullen, Dr J. M. Cullen and Dr R. A. Hinde for reading and criticising the manuscript.

II. An Attempt at a Rationale of the Taxonomic and Evolutionary Assessment of Behaviour Characters

Our comparative studies had two aims: (1) a mere description and listing of similarities and differences so as to arrive at a natural classification and to check the classifications given so far, and beyond this (2) an assessment of how and why these similarities and differences may have developed phylogenetically.

In the course of our studies it became gradually clear that much of the reasoning applied by us had often been vague, to a great extent intuitive, and less sophisticated than that which is now being developed in taxonomy (Cain & Harrison, **15**). Partly because of this, and partly because behaviour studies meet with some difficulties not found in morphological studies, a critical examination and re-appraisal of our methods seemed necessary. For practical reasons I will discuss some general aspects of these methods in anticipation of the results: some readers might prefer to read the factual chapters first.

In comparison of behaviour, as in morphological comparison, similarities and differences in characters are described. For this purpose species-specific characters are singled out, i.e. properties which are relatively constant throughout the species[1] either in all members, or in males, females, or a certain ontogenetic stage, and which differ between species or larger taxa.

In describing differences in characters, it is of interest to know to what extent these differences are phenotypic, and to what extent, if at all, genotypic. This is particularly important in behaviour studies, since behaviour can in so many ways be moulded by the environment.

Evidence on this point is still relatively fragmentary. Without entering into a detailed discussion of behaviour ontogeny, it can be said that, on the whole, movements of the 'fixed pattern' type (*Erbkoordinationen*, Lorenz, **44**) show a high degree of ontogenetic stability in the sense of resistance against environmental modification, the song patterns of some song birds (Thorpe, **73**) being a relatively rare exception. The scattered reports of development of motor patterns in animals raised in abnormal environments particularly show that the type of movement on which the present study is mainly based viz. displays, is highly environment-resistant (Goethe, **25**; Heinroth, **28**; Lorenz, **44**; Sauer, **69**). The situation is different with regard to other parts of the behaviour machinery. The selective responsiveness to specific releasing stimuli for instance is often subject to conditioning. Thus Bergman (**10**) showed that Turnstones hatched under Redshanks ignored the alarm calls of adults of their own species and soon responded to their foster parents. On the other hand, Redshank chicks hatched under Turnstones responded to the call notes of their own species; later they became conditioned to those of their foster parents. The specific responsiveness to some stimuli releasing food begging in young Herring Gulls (Goethe, **26**) and Black-headed Gulls (R. and U. Weidmann, **91**) is the same in 'naïve' chicks as in chicks raised by the parents, though it can be changed at a later stage by conditioning (E. Cullen, **18**).

Other aspects of behaviour machinery have also been shown to be

[1] For the present study, most gulls were studied in one or a few colonies only, and our facts may not always be representative in detail of all populations of each species.

environment-resistant: thus the difference in mobility between chicks of the Kittiwake and of the Black-headed Gull is truly innate (E. Cullen, **18**).

While for our present purpose the characters compared can safely be assumed to be valid from this point of view, the need for more precise ontogenetic studies is obvious.

The variety of behaviour characters observed in gulls allowed a considerable widening of the range of taxonomically useful characters. This allowed a better assessment of their taxonomic affinity. Since all gulls so far studied are very similar in most of their behaviour their close taxonomic affinity was thus confirmed.

For phylogenetic interpretation, our method amounted, in a nutshell, to the application in detail of the same method by which, e.g. whales and bats are found to be mammals and thus to be more closely related to each other than whales are to fish and bats to birds. While for purely taxonomic purposes this is concluded from the fact that whales and bats share more characters with other mammals than with fish and birds respectively, phylogenetic interpretation depends on 'weighting' of the various characters. This implies a refinement, through functional interpretation, in the crude use of sheer numbers of common or different characters. A study of function, or biological significance, can reveal the adaptive significance of (a) differences in single or a few characters between otherwise very similar species and thus strengthen the conclusion that such species are of common descent and have undergone recent adaptive radiation and (b) similarities of single or a few characters in otherwise different forms, and thus strengthen the conclusion that these similarities are due to convergence. The gulls being a monophyletic group, the present study is mainly concerned with adaptive divergence; but convergent traits shared with otherwise very different groups (analogous to the fish-like external shape of whales and fish) will also be briefly mentioned.

The weighting of characters on the basis of functional study has been done in the following way. Whenever, in a species which on total number of common characters is judged to be closely related to other species, the differences can be correlated to different functions and can thus be considered to be recently acquired, these differences can so to speak be 'peeled off', and a core of characters revealed which are either different for (as yet) unknown reasons, or which are shared with the other species. If such cores are more similar between species than their total complexes of characters, their affinity is even more probable than before functional studies were undertaken; if the cores turn out to be more different than the total complexes, common descent of the species compared becomes doubtful. Apart from

29

and beyond this, such functional studies allow conclusions about the particular *types* of selection pressure which must have been at work in adaptive radiation. Thus E. Cullen's study of the Kittiwake revealed that many behaviour (and morphological) peculiarities of this species have to do with its specialisation as a cliff breeder; the most striking differences between the various Galapagos finches were found by Lack (**40**) to be related to differences in feeding habits. When the Kittiwake is stripped of those peculiarities which can be recognised as cliff-breeding adaptations or their corrollaries, its 'core' of characters so revealed is strikingly similar to those of most other gulls. At the same time, the nature of its specialisations is better understood.

Since description precedes functional interpretation, great care is needed in the naming of the characters. In the present study, which concentrates on displays, naming has been done in the following way. Within each species, a number of displays have been distinguished. This is possible in spite of the fact that each posture shows considerable variation, because there is, in most cases, a clear discontinuity between the displays. Further, each display, or rather the range of forms in which the same display can appear, is strikingly constant within the species. Third, in spite of the constant differences between species, there is, for each display, a degree of similarity between species often as striking as the morphological similarities.

Because the displays are at first described with regards to their formal appearance only, the names should be descriptive without any reference to either function or causation, for uncertainties in interpretation may lead to confusing changes of names as these interpretations change. The similarity between species allows the application of the same name to similar displays in different species. However, this is justified only if and when the same name can be assumed to indicate that the displays are not convergent or accidentally similar; application of the same name ought to indicate common descent, or perhaps parallel descent in many species. Strictly speaking therefore (since common descent is a matter of conjecture) the safest course would be to start by applying each name to one display of one species only. As the study proceeds, this procedure (which, while consistent, is of course pedantic and clumsy) can be simplified by allotting common names. This causes no great difficulty where interspecific differences are very small, such as in the Oblique, or in Facing Away, but with greater dissimilarity, such as between the Kittiwake's Bow-and-Moan and the Herring Gull's Mew Call, the uncertainty may be such as to warrant the use of different names, and thus avoiding prejudging the issue.

For these reasons it is important to be clear about the criteria used in the assignment of common names. Convergences or accidental similarity between displays is considered to be improbable when (1) their interspecific formal similarity is great (particularly when the displays are complex and 'improbable'); (2) their distribution through the group is wide and continuous; (3) they are, in addition, found to have a similar motivation, and (4) a similar function throughout the group; and, finally (5) when they can be considered to have had the same origin. This latter aspect is particularly important in displays, for most, if not all are now recognised as being 'derived' movements (Lorenz, **43**; Daanje, **21**; Tinbergen, **77**).

In practice, various difficulties arise. First, formal similarity, which per definition is not identity, can be great or small. Thus while there is not much doubt that the Upright of the Herring Gull and that of the Black-headed Gull deserve a common name, this is less certain of the Head Flagging (Facing Away) of the Black-headed Gull and the Facing Away of the Kittiwake.

When the criterion of formal similarity cannot be applied without hesitation, but there are indications that function and causation may be very similar, such as is the case with the Vertical of the Little Gull and the Forward of the Black-headed Gull (which are both shown as the second posture in the meeting ceremony), uncertainty is best expressed by applying different names. This may or may not be temporary. Conversely, although there is a certain similarity between the Forward of the Black-headed Gull and 'Jabbing' of the Kittiwake they appear in different circumstances and therefore must have different function and causation. For these reasons, naming is provisional and tentative, and may be subject to subsequent changes.

As to (5), movements such as the Upright of which the origin is clear, and Choking, about whose origin a relatively clear picture is gradually emerging, can be named with more confidence than the Mew Call Posture, about the origin of which there is as yet not much of a clue.

Apart from all these considerations a difficulty arises, as in comparative anatomy, whenever a posture seems to have been fused with other elements in some species while not in others. Thus the Oblique-cum-Long-Call of the Herring Gull is really a complex of three movements; while the Oblique-proper is very similar to that of many other species, it is preceded by a bending-down movement which is not found in the Kittiwake and the Black-headed Gull, and it contains the 'Throwback' which is again not found in the latter two species, whereas just this element is still more conspicuous in the Common Gull (Fig. 3). Similarly the Kittiwake's Bow-and-Moan contains one element which is very similar to the Mew Call Postures of other species, but the Bow may be an added element (Fig. 8). The Long Call sequence of the Ivory Gull is peculiar in that the Oblique is preceded by a posture very much like the Forward of the Black-headed Gull (Fig. 4).

This is one reason why the naming is no more than a crude approximation; the situation is rather similar to that in the comparison of the skulls of various vertebrates, and can be dealt with in the same manner.

Differences between postures receiving the same name may also be due to 'ritualisation' (see chapters VI and VII) of the movement itself. As will be seen, the Choking of the Kittiwakes is different from that of other gulls in that the mouth is usually wide open (Fig. 7); this no doubt has to do with the fact that the inside of the mouth is a very bright orange, a device which presumably emphasises the conspicuousness of this posture. A similar functional explanation must probably be given to the spreading of the wings by the Great Skua when performing the Oblique-cum-Long-Call, this shows off the bright white patches on the wings. Without committing myself in the question which came first, the change in structure or the change of movement, I suggest that the correlation is functional.

Thus the evolutionary interpretation of these comparative behaviour data is beset with difficulties. Some of these cannot be removed—we cannot hope for direct evidence on the behaviour of the gulls of the past. This means that our reasoning has to be 'probabilistic' (to use a term applied by Thieme (**72**), for the procedure of comparative linguistic studies); by increased precision of description, by extension of the number of species studied and of the characters used, by study of functional aspects, and by continuous scanning for inconsistencies or contradictions in the total set of conclusions the probability of the overall picture can be judged with increasing confidence. If the premises are false, inconsistencies are bound to appear; if uncertainties and inconsistencies tend to be reduced by more intensive and extensive study the premises will be strengthened. Even so, the gaps between present species may often be too wide to allow a full phylogenetic interpretation.

III. Description of the Agonistic and Pair Formation Displays

This chapter will list and briefly describe the main displays found in all or most of the species so far studied. Before doing so I have to anticipate some of the results in order to justify the terminology. (1) It was found that, in spite of great intraspecific and even intra-individual variation, there is, within a species, a striking discontinuity between displays. Thus, in spite of the various forms which the Oblique, the Choking and the Forward may assume (the causes of which will be discussed in chapter v), it is possible to state in the great majority of observations which display the bird is performing. As an illustration, Fig. 1 shows the result of the analysis of three cine film shots of a male Black-headed Gull adopting the Oblique and the Forward as a response to flying birds. The number of frames showing either posture or the transitions from one to the other were counted. It will be seen that in comparison with the time taken by each posture, the transition was extremely short, the average of six transitions

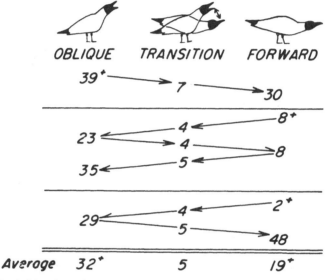

Fig. 1. Number of frames in three cine film shots of the Oblique, the Forward and the transitions between them. Black-headed Gull. Speed 24 frames/second.

being about one-fifth of a second, during which the bird's head moved over a distance of at least four inches. (2) Each display is constant within the species; thus all individuals of the same sex and age have roughly the same range of Obliques, Choking postures, Forwards, etc. (3) There are relatively constant differences between species: for instance the Oblique of the Kittiwake is consistently different from that of the Herring Gull. (4) In spite of the differences between species, there is, for each display, a degree of similarity between species as striking as the morphological similarities.

33

Fig. 2. The upright posture. a. Aggressive Upright of Herring Gull;
b. Intimidated Upright of Herring Gull; c. Anxiety Upright of Herring
Gull; d. Aggressive Upright of Hartlaub's Gull; e. Anxiety Upright
of Hartlaub's Gull; f. Aggressive Upright of Common Gull; g. Aggressive Upright of Black-headed Gull; (d. and e. after Tinbergen &
Broekhuysen, 85; f. after Weidmann, 90).

The following list includes what could be called the elementary displays, not the species-specific display sequences which will be discussed later.

1. *The Upright* (Fig. 2)   The neck is stretched more or less vertically. The bill can be horizontal, or can be pointed slightly down or up. The carpal joints can be raised, or can be hidden in the supporting feathers of the flank as in the rest position. This display can be seen in boundary disputes or in pair formation sequences. During or after a charge, the wings may be fully raised and spread. No sound is made. The bird may be standing still, it may be walking up towards another bird; it may turn its side or even its back towards the other.

As will be described later, the posture can range between three extremes: the Aggressive Upright (Fig. 2a), the Intimidated Upright (Fig. 2b)[1] and the Anxiety Upright (Fig. 2c). In pair formation sequences of the Black-headed Gull an Upright is assumed in which the bill points down more than observed in Aggressive Uprights (Fig. 14).

The Upright is found in all species observed so far, except the Kittiwake, which has only the Anxiety Upright.

2. *The Oblique-cum-Long-Call* (Fig. 3)   The neck is stretched obliquely forward and upward, with considerable inter- and intra-specific variations in the vertical plane. The carpal joints are usually raised. In this position the bird utters a loud, long-drawn call with the bill wide open. In most species this call is repeated a number of times. The Oblique-cum-Long-Call may be seen in hostile encounters and in pair formation sequences. It is found in all species; the call itself is very different from one species to another: melodious in the Herring Gull, Western Gull, Iceland Gull, Glaucous-winged Gull; hoarse in the Lesser Black-backed, the Greater Black-backed and the Glaucous Gull; it has a crowing quality in the Black-headed Gull and Hartlaub's Gull; it is very high-pitched in the Common Gull; it is the Kittiwake's three-or-four-syllable 'kittiwaak'-call, etc.

In the Herring Gull there is a series of postures (Fig. 4b): in a complete Long Call performance the head is first pointed obliquely forward and slightly up, in which position one, usually hoarse, call is given. Then the head is jerked deep down, and one or two muffled, high-pitched calls are given; finally it is thrown up with a jerk (the 'Throwback') and a series of loud calls begins; with each call the head is lowered a little so that the bird ends in an almost horizontal

---

[1] Tinbergen (**79**) and Moynihan (**58**) do not distinguish between the Anxiety Upright and the Intimidated Upright and apply the name Anxiety Upright to both.

position. The Western and the Glaucous-winged Gull lack the Throwback (Fig. 4e); in the Common Gull it is extremely pronounced (Fig. 4a); the Ring-Billed Gull throws the head up with each call in the third phase; in the Laughing Gull a series of Throwbacks follows phase 3; although in both these species the Throwback is rather

Fig. 3. The Oblique posture. a. Herring Gull; b. Little Gull; c. Black-headed Gull, Aggressive Oblique; d. Kittiwake; e. Hartlaub's Gull; f. Intimidated Oblique of Black-headed Gull; (b. after Moynihan, **58**; e. after Tinbergen & Broekhuysen, **85**).

similar to the Head Tossing of other species, its relationship with this display will be left undecided here. In the Ivory Gull the Oblique, rather similar to that of the Black-headed Gull, is regularly preceded by a posture very similar to the Forward (Fig. 4d).

In what seem to be incomplete, low-intensity performances a bird may give 'Long Call Notes', single calls very similar to the Long

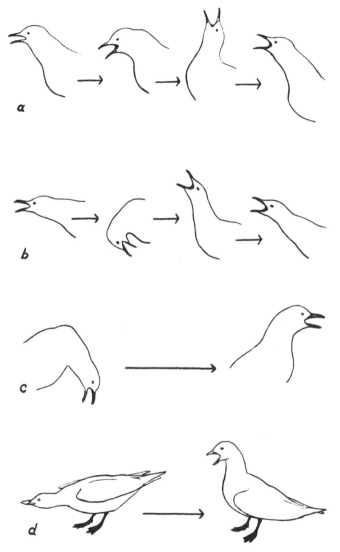

Fig. 4. Oblique-Long Call sequences to be read from left to right.
a. Common Gull; b. Herring Gull; c. Western Gull; d. Ivory Gull;
(a. after Weidmann, 90; d. after Bateson & Plowright, unpubl.).

Call but not repeated, and given in a position more or less intermediate between the Oblique and the Upright.

3. *The Swoop-and-Soar* The Black-headed Gull and related species
often pursue other birds in the air and swoop down on them.
Usually they glide over the other birds and, with the wings fully
stretched and lifted in a V-posture, soar steeply upwards until they
almost stall. During the pursuit and the swoop the bird utters either
the Long Call or the Attack Call (see under 5). It seems likely that the
aerial performances of the Little Gull (L. Tinbergen, **74**; Moynihan,
58) and the large gulls are the same as the Swoop-and-Soar, less
ritualised in the Herring Gull group and more so in the Little Gull.

4. *The Forward* (Fig. 5) The body is about horizontal and the neck
is stretched to a varying extent. The bill is pointed horizontally forward or slightly up; it can be slightly opened or closed. The carpal
joints are often raised. It may be given without a sound, but more
often a muffled version of the Long Call note is given. In this posture,
which is given in hostile encounters and in pair formation sequences,
a bird can face another bird, stand parallel to it, or take up an intermediate position; it can even face away from the other bird. It is very
common in the hooded gulls. It looks very much like a derivate of the
Oblique, with which it is often linked through intermediates. In the
Little Gull (Moynihan, **58**) the neck is stretched vertically up (Fig.
5d), and the posture is thus reminiscent of the 'Erect' of terns (J. M.
Cullen, unpubl.) (Fig. 16), but it seems to be more closely related to
the Forward since a muffled Long Call is given, while terns in the
Erect are silent. We will later see that in motivation and function it
also resembles the Forward.

5. *Jabbing* (Fig. 6) This is one of a series of fast thrusts, usually
with open bill, in the direction of another bird. It may result in an
actual peck, but most often birds stop just short of touching the other
bird. The plumage of the occiput and the neck is often raised. No
sound is given. In Jabbing, a bird invariably faces another bird. It is
common in the Kittiwake, and occurs occasionally in other species.
The 'Gakkering' of various gulls (Moynihan, **58**) and the Sandwich
Tern (Van den Assem, **4**) seems to be the same as the 'Attack Call' of
the Black-headed Gull (Moynihan, **58**) which is not discussed there.
Originally (even as recently as 1958) I lumped the Forward and the
Jabbing together, but I now believe that Moynihan is right in separat-
ing the two, mainly because of the facts that in the Forward a form of
the Long Call is given; that the Jabbing is a clear pecking movement,

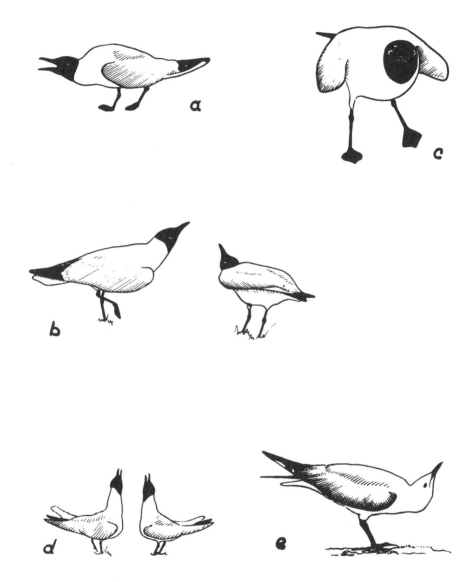

Fig. 5. The Forward posture. a., b. and c. Black-headed Gull; d. The Vertical of the Little Gull; e. Hartlaub's Gull; (d. after Moynihan, 58; e. after Tinbergen & Broekhuysen, 85).

Fig. 6. Jabbing of the Kittiwake.

not a posture; and that in Jabbing part of the plumage is raised.

6. *Choking* (Fig. 7)   In this posture, the bird squats and bends forward. The tongue bone is usually lowered, the neck is held in an S-bend, and the bill is pointing down. In this position the head makes rapid downward movements, usually however without touching the ground. The carpal joints are often raised, and the wings may even be raised and spread, and kept stationary for seconds. A muffled, rhythmical sound is given which may or may not be in time with the pecking movements. The breast is 'heaving' strongly, particularly in the large gulls. Often the lateral ventral feathers are raised. The bird may be facing another bird, or face away from it, or take up an inter-

Fig. 7. Choking. a. Herring Gull; b. Common Gull; c. and d. Black-headed Gull; e. Kittiwake; (b. after Weidmann, **90**).

40

mediate orientation. Choking seems to occur in most if not all species of gulls. The more distantly related Great Skua does not have it (Perdeck, **66a**) and neither seem the other skuas. Nor has it been reported from terns, except the Brown and the Black Noddy (J. M. Cullen, **20**) which have a display very similar to it. Fig. 7 illustrates some of the inter- as well as the intraspecific variations. Choking occurs in hostile clashes as well as in pair formation and nest-showing sequences.

Fig. 8. The Mew Call posture. a. Lesser Black-headed Gull; b. The Bow-and-Moan of the Kittiwake; c. Common Gull; (c. after Weidmann, **90**).

7. *Upward Choking*   This movement seems to be peculiar to the Kittiwake. At the end of the pair formation and meeting sequence Choking gradually changes into it; the bill is pointed more and more up, and the mouth, which in Choking is wide open, is gradually closed (Fig. 14).

8. *The Mew Call* (Fig. 8)   This is a long-drawn, often plaintive call, given with the bill wide open, or, in the Black-headed Gull (where Kirkman, **38,** calls it the Crooning Call) closed, and usually in a forward-bent posture. The character of the sound is always very different from that of the Long Call. The carpal joints are often raised. The bird may run up towards another bird in this posture or it may stand still, or walk parallel to another bird. The Herring Gull and its relatives bend the neck in a characteristic way (Fig. 8a). So

41

does the Kittiwake in the 'Bow-and-Moan' (Fig. 8b), part of which seems to be the Mew Call posture. The Mew Call is uttered in hostile situations, during pair formation, as part of the nest-showing ceremony, and during incubation and caring of the chicks.

9. *Pecking-into-the-Ground* (Fig. 9)   This is done in territorial clashes by the members of the Herring Gull group, by the Kittiwake and, in a less violent way, by the hooded gulls. The carpal joints are usually raised; the wings may even be fully spread in the Herring Gull group. The latter species follow it up with picking up or pulling out of nest material, which is then either dropped or thrown sideways as in

Fig. 9. Grass Pulling, Lesser Black-backed Gull.

building. This is why I called the movement (79) 'Grass Pulling' and interpreted it as displacement nest building. I will discuss its relation with nest building (and other movements) in chapter VI. The Ring-billed Gull and some hooded gulls are also reported to pick up plant material after pecking down. The Black-headed Gull, males of which may peck at the ground when visited by a female, often follows it up with bill wiping.

10. *Preening* In certain ceremonies, such as the pair formation sequence of the Black-headed Gull, a certain type of preening is regularly seen. The bird may just make one 'emphatic' sweep under the carpal joints, or sometimes in the scapulars. It is less regularly seen in the Herring Gull group, and it seems to be absent in the Kittiwake. We probably have given it insufficient attention, but it is doubtful whether it has any signal value.

Fig. 10. The Hunched posture. a. Fully fledged young Herring Gull;
b. Black-headed Gull; c. Kittiwake (female, left).

11. *The Hunched* ('*Submissive Posture*') (Fig. 10)   This resembles the
Forward but in its extreme form the back is hunched and the neck
is entirely withdrawn. The carpal joints are not raised. The bird may
be silent, or (usually when the posture is shown by grown-up young)
utter a soft, high-pitched, drawn-out sound. Observed in all species. It
is often shown by females when near their males (when they may com-
bine it with Head Tossing), and by grown-up young when near adults.

12. *Head Tossing* (Fig. 11)   The head is tossed up with a quick
movement so that in extreme cases the bill may point vertically upward for a moment. The bill may be opened slightly (Herring Gull
group) or wide (Black-headed Gull group) or may be kept shut
(Kittiwake). It may be done once, or may be repeated with longer

Fig. 11. Head Tossing, Herring Gull. The two extreme head positions
are drawn in.

43

(Herring Gull group, Black-headed Gull group) or shorter (Kittiwake) intervals. With each toss, a soft call is given, which is very melodious in the Herring Gull group, high-pitched and shrill in the Black-headed Gull group, and very high and squeaking in the Kittiwake. Usually Head Tossing is done in the Hunched, but in the early stages of pair formation a female may do it in the Anxiety Upright. It occurs prior to copulation (when both sexes perform it) and prior to feeding (when it is done by the large young, and also by the female when she solicits courtship feeding in the male).

13. *Facing Away, or Head Flagging* (Fig. 12)   This movement is conspicuous in hostile encounters in the Kittiwake and in the Black-headed Gull group; in the latter group it is a very regular feature of the pair formation sequence; in the Herring Gull group (where I overlooked it until E. and J. M. Cullen pointed it out to me) it is less regularly but yet quite often seen in the same situation, but the movement is much less striking than in the Black-headed Gull. The head is turned away with a jerk (most pronounced in the pair formation sequence of the Black-headed Gull group); in both the Herring Gull and in the Black-headed Gull group the movement is in the horizontal plane, but in the Kittiwake a down component is present. Whether or not the Little Gull has the same movement is uncertain; Moynihan (58) has seen inconspicuous side to side movements of the head. The movement is usually done in the Upright (either in the Anxiety Upright as in the Herring Gull group, or in a posture very similar to the Aggressive Upright as in the Black-headed Gull group), but it can be shown in any other posture.

14. *The Alarm Call*   All species have a short staccato call which is uttered as a response to a predator and, in rare cases, in intraspecific clashes, when it is given by a thoroughly beaten bird. The Black-headed Gull utters the calls singly, with only a slight tendency to rhythmic repetition. In the Herring Gull group it is usually given in bursts of three to six calls. The Kittiwake's alarm call is heard only upon strong provocation; it is rhythmic but the single calls are less staccato. The Little Gull's alarm call is described by Moynihan as an extremely sharp 'truk', which may be uttered in series.

IV. Functions of the Displays

In comparison with other, more 'straightforward' behaviour patterns such as feeding, or escape from predators in the individual sphere, and copulation in the inter-individual sphere, displays offer special

Fig. 12. Facing Away. a. Kittiwake (the lower bird, who is an intruder on the nest of the Long-calling pair); b. Lesser Black-backed Gull, female (left) and male; c. Kittiwake (the lower bird; the other bird jabs); d. Common Gull (the bird on left; the other adopts the Aggressive Upright); e. Young Kittiwake; f. Hartlaub's Gull; female (left) and male; g. Black-headed Gull; female (front) and male; (d. after Weidmann, **90**; e. after E. Cullen, **18**; f. after Tinbergen & Broekhuysen, **85**).

difficulties to an understanding of biological function and of causation. Feeding leads to the intake of food, which is known to be indispensable for the maintenance of the individual, and its overall function is therefore clear—present-day research concerns much more detailed problems concerning the exact functions of food. Knowing what happens when a predator overtakes its prey, the overall function of escape is also clear, although again many finer details of anti-predator behaviour are still ill understood. Equally the function of copulation is known. As to causation, we know that starvation and stimuli provided by appropriate food cause feeding behaviour, and we know that escape behaviour is elicited by stimuli provided by the usual predators. Copulation is known in many cases to depend on the endocrine state and on stimuli provided by the partner.

However neither the functions nor the causation of displays were obvious at first glance, and until quite recently they were only vaguely understood. Although much progress has been made in the last two decades, their study is still very far behind that of most other behaviour patterns. This is partly due to the fact that displays are varied and not easily described, partly to the circumstance that their effect is less direct than that of other patterns (since they act through the sense organs of other individuals), and partly to the fact that work on internal and external causal agents has been extremely patchy.

However, mainly through the work of Lorenz (**43**) and others, it is now clear that the overall function of displays is that of 'releasers' (Lorenz) or signalling devices, and as such they are a distinct class of effector organs, functioning in inter-individual relationships. For a recent summary of the observational and experimental evidence I refer to Tinbergen (**80**). Even now our knowledge is both fragmentary and superficial.

We have given attention to problems of function both for purposes of a better understanding of the social organisation of the breeding communities and of the adaptive radiation of the displays within the group.

METHODS

While most of the displays of gulls are undoubtedly releasers, experimental evidence of this concerning the displays discussed in this paper is practically absent. Pilot tests with the aid of 'dummies' were repeatedly made but were defeated so far by practical difficulties: non-feathered dummies evoked intense fear, and feathered dummies were usually attacked fiercely, and were completely demolished and useless after a few tests. Since our main interest was focussed on other problems, experiments of this type were not continued.

However, the study of unplanned, or 'natural' experiments has enabled us to make fair guesses as to the functions of most of the gulls' displays, and have led to some hypothesis which seem worth checking. While such natural experiments do not as a rule provide information about the part of the total display which is effective (e.g. whether sound, movement, posture, or colour is the main agent) they do, if observed with care and with an eye for possible flaws and for 'natural controls', and if repeated often enough, tell us a great deal about the functions of the displays as wholes. All the same, the conclusions drawn in this section must be regarded as tentative.

While it would be beyond the scope of this paper to present detailed evidence on each display discussed, a few examples of the type of observation used can be given:

(1) March 10, 1957. Ravenglass. Two Black-headed Gulls are observed feeding on a gravel bank in the estuary. They are obviously a mated pair. While feeding they wander apart, the male feeding on one side of the bank, the female on the other. Thus the birds are usually separated by the bank, which is about 3 ft high. In the course of 70 minutes the male performs the Oblique-cum-Long-Call 16 times while the female is out of sight. These calls are responses to other birds approaching him. In 13 out of these 16 cases the female runs up to the male immediately when he calls. (On the three occasions when she failed to do so she was herself engaged in a squabble with another bird.) Such observations, compared with the absence of a response from the female when the male was not calling, can be made repeatedly if the observer is alert to the possibilities; they allow one to conclude that the Oblique-cum-Long-Call makes the female approach the male. In this particular case, it was obviously the sound itself and not visual stimuli which attracted her. The observations do not reveal anything about the effect of the posture and call on other males.

(2) April 28, 1955. The Inner Farne (E. Cullen). A male Kittiwake (No. 20) 'sings' (Chokes) on his ledge; a female visits him. They perform the meeting ceremony (see chapter v); the female shows Head Tossing, standing with her side to the male. Approximately 1 minute after her arrival the male pecks at her; she then faces away and turns her beak down. The male at once stops his attack, and just looks at her. After a while she lifts her head and begins to turn her face towards him; he now grabs her cheek, obviously aiming for her beak as it becomes visible to him. She leaves the ledge.

This type of observation was done a great number of times in the course of four breeding seasons, and there is no doubt from the scores that Facing Away stops, or something merely reduces, the male's attack. The opposite (Facing Away stimulating the attack) was never observed.

(3) April 14, 1958. Ravenglass. A male Black-headed Gull is walking on his territory, gathering straws which he carries to his nest site. While doing so he repeatedly walks in the direction of a group of five birds which are standing and preening on a bare patch of sand just outside his territory; he is always absorbed in looking for and collecting straws. After his building bout has subsided, he stands near his nest, preening. Suddenly he adopts the Aggressive Upright, and walks up to the five gulls. When he is still about 10 ft away the five gulls adopt the Anxiety Upright, and begin to walk away. When they have gone about 12 ft the owner of the territory stops the pursuit.

Such observations, in which birds move away when a territory owner approaches them in the Aggressive Upright although they ignore him when he approaches them in another way, have been made many times.

While more detailed evidence on the functions of the various displays will be given in monographs of the separate species which are being prepared (e.g. Manley, **47**), some general features will be discussed here.

The postures listed are used in agonistic (Scott and Fredericson, **70**) and related situations such as pair formation (see chapter v). According to their functions they can be divided into two groups. Group 1 I will call spacing-out, or distance-increasing displays; group 2 distance-reducing displays.

Group 1 comprises the Aggressive Upright, the Oblique-cum-Long-Call, the Forward, Jabbing, the Mew Call, Choking, and Pecking-into-the-Ground. They promote spacing-out of breeding pairs within the colony in two different ways, depending upon the situation in which they are shown. When shown by a territory owner in his territory, they have an intimidating effect on intruders or near-intruders; it is this effect which had led us to call them 'threat displays'. They tend to make such outside birds either move away, or stop their advance. The displays may however also be shown by intruders or near-intruders, and then their effect on neighbouring territory owners is just the opposite: they provoke the tendency to attack (see Tinbergen, **79**). This attack in its turn tends to make the intruders move away or stop an advance, and thus eventually the displays promote spacing-out in this situation as well. Spacing-out therefore depends on two tendencies, present in each bird, particularly the males: that to attack a rival when it intrudes into one's territory, and that to withdraw from a rival when it is met with outside one's territory (see Lack, 41; Tinbergen, **83**). Both tendencies are evoked by the opponent's agonistic posturing; which of the two is strongest depends on the locality with reference to the reacting bird's territory.

All species of gulls have more than one display in this category, and it is of course necessary to ask why this should be so. There seem to be two reasons. Lorenz (**44**, **45**), Marler (**49**) and others have pointed out that displays may have to be distinct in order to be unambiguous; intraspecific unambiguity reducing the danger of misunderstandings or inappropriate responses. This, while explaining why for instance a distance-reducing display should be different from a spacing-out display, does not fully explain the existence of more than one spacing-out display as long as these have not been shown to have different functions, for unambiguity is necessary only between displays with different functions. I suggest that each display is an adaptation to one of a limited number of frequently occurring situations, and that this is also the explanation of the fact that each species has only a few agonistic postures—of course, this limitation to a small number of displays is just as much of a problem. If (1) there is a limited number of frequently occurring situations which have to be dealt with, and if (2) each posture has evolved as an unambiguous way to achieve this, then the possession of more than one display and yet not more than a few would be adaptive, and this would explain the need for unambiguity.

Without claiming to understand the exact function of each of the displays, I believe that a distinction must be made between relatively more 'offensive' and relatively more 'defensive' displays. For instance, the Aggressive Upright and Grass Pulling in Herring Gulls are understood by other gulls as meaning 'He is ready to attack', whereas the Forward of the Black-headed Gull and especially the Choking of all species rather convey the message: 'Don't attack—if he is attacked, he will fight back'. In other words, the former displays drive intruders away, the latter stop birds from trespassing. We will see in chapter v that the motivation of these displays can actually be described in these ways.

Further, in each of these sets of two postures there seems to be one which has a milder effect than the other. Grass Pulling seems to be more frightening than the Aggressive Upright, and Choking has a stronger containing effect than the Forward. The justification for the existence of two intensity types in both the more offensive and the more defensive displays may be that there are, again roughly speaking, two distinct types of intruders: the casual, accidental intruder

49

who alights without any serious intention of claiming a territory, and the would-be settler, or the already established neighbour who tries to expand his territory. The first type flees at the slightest warning; the latter is usually prepared to stand his ground tenaciously, and requires 'strong language'. A strange bird can usually be classified as either one or the other; intermediates are relatively rare.

Finally, it is striking that most species have one extremely loud call—the Long Call. This may be because a signal perceived from a distance might be useful in preventing birds from alighting at all. (It must be admitted that the Kittiwake does not seem to fit in here; it uses the Choking rather than the Long Call as a distance signal.)

Thus my suggestion is that there might be an advantage in having five different threat displays. One is used as a long-distance threat (and song, see below); then there is one aggressive and one defensive display effective with accidental intruders and near-intruders respectively; and one aggressive and one defensive display effective with determined intruders and near-intruders. Yet this can not be the whole story; apart from the discrepancy pointed out in the Kittiwake the Mew Call does not seem to fit in a clearly defined pigeonhole, nor is the function of each display exactly the same in all species. Clearly more purposeful and systematic investigations of the functions are required.

So far, I have been speaking of the effect in purely agonistic situations where the displaying bird deals with males. Several of the displays however have an effect on the female as well. This effect is often entirely different. In most species, the Oblique-cum-Long-Call, apart from intimidating males, attracts unmated females. Because of this dual effect the name 'song' can be applied to characterise its function. In the Kittiwake, the male's Choking functions as song. The Mew call also attracts females in many species; it is effective particularly at a later stage of pair formation when the birds become interested in a nest site. Choking, apart from acting as song in the Kittiwake, is used by many if not all other species to call the female to the nest or prospective nest site ('Friendly Choking').

The conclusion that displays of group 1 promote spacing-out has received striking confirmation from two sets of observations in which the spacing-out (which is essential to the smooth functioning of colony life) failed due to the absence of these displays.

(1) In the Herring Gull colony on the Frisian Island of Terschelling I often observed immature birds preying upon chicks. As I described before (79), such birds walked up very slowly to a nest with one or more chicks covered by the parent. Even when the male was present on the territory (the female covering the brood), he would not

succeed in driving the intruder off, in fact he did not even attack it. He would walk up towards it in the Aggressive Upright, but the only response of the intruder was a slight stretching of the neck, and taking one or two steps aside. It neither postured, nor showed any signs of fleeing or of attacking in return. (This is in itself an interesting demonstration of the fact that the tendency to attack and that to flee from another male are both greatly increased when birds come into reproductive condition—these immature birds did not show reproductive behaviour of any kind and were attracted to the colony as to a feeding ground). The absence of any display by the intruder seemed to paralyse the male. Every few seconds the intruder would take a few small steps towards the nest, until it finally stood, completely unmolested, less than a foot away. It then waited until a chick showed itself, grabbed it with lightning speed, and flew off with it in its bill, the parents (provoked at last into action) pursuing it in vain.

(2) In several Black-headed Gull colonies G. Manley and I myself observed many males which tried to mate with females not their mates. Such males seemed possessed by one urge only: copulation. They walked over territories practically unmolested, and again the most marked feature of their behaviour was the almost complete absence of signs of fear and of aggression. They did stretch their necks, but in a way similar to that of any male prior to copulation, and now and then they flew up, and tried to alight on a female from above. They were often prevented from actually mounting by the females' attacking them, but this was the only overt avoidance they showed, and most of them persisted for minutes on end. Even when the mates of such females were present, they behaved as the territory owners in the first example; they might adopt low intensity threat postures but failed to follow up with an attack.

It is the remarkable contrast between these observations (which were both made a number of times) and the observed effects of threat displays when shown, which strengthens the conclusion that threat postures are important, in fact are the main spacing-out devices; if, for whatever reason, they are absent, spacing-out breaks down, in both cases with highly disruptive effects.

Group 2, in which I place the Hunched, Head Tossing, Facing Away, and Upward Choking, achieve the distance-reducing effect in two different ways. The Hunched, Facing Away, and possibly Upward Choking, seem to directly inhibit the tendency to attack or to withdraw, rather than, as threat postures do, to stimulate one of these two tendencies and thereby overriding the other. For instance, Dr E. Cullen has numerous records of the type of the Kittiwake observation mentioned above in which an aggressive bird, meeting a

Fig. 13. Three stages of the 'meeting ceremony' of the Black-headed Gull. Top. male 'long-calling'. Centre: male (left) and female in parallel, bill-up Forward posture. Next page: male (right) and female 'Facing Away'.

bird which faces away, calms down. For this reason we have suggested (Tinbergen and Moynihan, **86**) the term 'appeasement posture' for this category. Head Tossing reduces spacing-out in a different way; it releases either courtship feeding or copulation (which of the two depends mainly on the state of the responding bird) and thus evokes a behaviour tendency which itself overrides both agonistic tendencies.

Reduction of distance plays a part in the social organisation in various situations, and in each situation where one of the postures of group 2 is shown the need for distance-reduction is clear. Thus in the Herring Gull group and in the hooded gulls, Facing Away is shown in pair formation sequences, and it then occurs at the end of the 'Meeting Ceremony' (Fig. 13). In this situation (as will be discussed more fully in the next chapter) it counteracts the effects of the earlier components of the sequence (which contain agonistic elements) and thus opens the door to sexual cooperation between the partners of a pair. In the Kittiwake it further occurs during fights and it is often shown by birds which settle on an already occupied nest. Such a bird is not particularly aggressive itself; its overriding tendency seems to be just to sit on the nest. It is however strongly attacked by the owners, and then it Faces Away; this clearly reduces and often totally inhibits the attack; the owners' aggressive tendencies are then vented either in vigorous Long Calling or in attacks upon each other; but as a rule hostile displays die down. In the Black-headed Gulls, Facing Away is also shown in fights; here it may be shown by the individual which is beaten, and it then seems to help in reducing the vigour of the attack. Rather surprisingly it is also often shown by an attacking bird, and then it is perhaps of advantage either in delaying the opponent's withdrawal or its counter-attack; we have no observations yet which throw light on this. As will be seen in the next chapter, this fact is also puzzling with regards to the motivation of Facing Away. In Kittiwakes finally the young Face Away when attacked by nest mates or by strange adults. The survival value of this is particularly clear, since it helps to prevent perilous fights on the narrow ledges. In all these examples the display allows the birds to be nearer to each other than they could otherwise be.

The Hunched Posture seems to have a slightly different effect. In 1953 I suggested that in the Herring Gull the effect of the Hunched, when adopted by large young, was to reduce aggression by the parents. R. Weidmann (89) however did not see many signs of aggression in Black-headed Gull parents, and she stresses the tendency of the parents to avoid large young, particularly when they are vigorously begging for food. I must admit that this avoidance tendency is more common than aggression in the Herring Gull as well, and it seems likely that the main function of the Hunched, of large young at any rate, is reduction of this avoidance tendency. This tendency need not be identified with fear (meant, of course as shorthand for a tendency to flee or to withdraw from an attacker).

It may well be that the Hunched and Facing Away are distance-reducing displays in slightly different senses; the Hunched prevents another bird from withdrawing rather than from attacking, Facing Away stops an attack rather than withdrawal. In chapter v I will discuss why spacing-out and distance-reducing displays are used in pair formation as well as in agonistic situations.

It must be stressed that this sketch of some of the functions of the displays is not only tentative, but also gives a much simplified picture. Nor did I discuss all the displays and all the functions. For instance the Mew Call serves in many species to attract the young prior to feeding them, but this was not discussed because this paper concentrates on the agonistic and pair formation displays. Nor did I give more than passing attention to the fact that a display may have different functions in different species; this aspect will be discussed in chapter VII. Finally I did not discuss in any detail the variety of forms in which each posture can appear even in one individual. We know more about the causation than about the functions of these variations and they will therefore be treated in chapter v, but it is worth mentioning here that there are indications that gulls 'understand' even some of these variations, i.e. respond to them in appropriate ways. Finally no mention was made of the possible changes of the functions in ontogeny; we simply do not know enough about this.

V. Causation

Previous studies of the causation of displays have drawn attention to the phenomenon of dual and multiple motivation (see Andrew, **3**; Hinde, **30**, **32**; Lorenz, **46b**; Marler, **48**; Morris, **55**; Tinbergen, **78**, **79**, **81**). In this they differ from many other types of behaviour, such as feeding, or nest building, or copulation, which can be said to be controlled by single or unitary motivation. Misunderstandings about the significance of the recognition of dual and multiple motivation (which must be considered the first step in the causal analysis of displays) have arisen partly through a relatively loose usage of terms in ethological analyses. This has been realised by several workers in this field, and in attempts to improve methods of analysis several of the authors mentioned have discussed the theoretical background, the formulation of the problems involved, and the methods best suited for their study. These motivational studies are admittedly still in a very early stage, and we are still far from understanding the displays of gulls. Yet some interesting though tentative conclusions begin to emerge. In this chapter I shall discuss some concepts and some methods used, and some of the results so far obtained.

The concepts used in this analysis, and dictated by its aims, are 'motivation' and 'tendency'. I use both as mere shorthand descriptions of the state of an animal as judged by the movements it makes. Of course this state is due to a complex of both internal and external conditions, the unravelling of which is normally the first in the analysis. However, for our present purpose (the analysis of the displays) one step has to be made even before this: splitting up the total motivation in two or three component motivations, each of which is unitary in the above sense and thus comparable as regards degree of complexity with feeding, etc. The unitary systems dealt with in this paper are attack, escape and sexual behaviour (in the widest sense, including pair formation behaviour). Of course, each of these subsystems offers problems of causation of its own, but the present study is not primarily concerned with the analysis of each of them separately, but with their effects when aroused simultaneously. To indicate the complex of conditions which make a bird attack, escape or mate, we need a term of the same type as 'motivation', but applying to one of these single systems only. In order to avoid the word 'drive' which is ambiguous because it has been used in different senses, I use the word 'tendency' (Hinde, **32**, Tinbergen, **82**). While this word again is nothing but a shorthand description of the state of an animal as indicated by the movements it makes, it is needed to describe the total motivation of an animal at any given moment in terms of the simultaneous arousal of two or more tendencies in the above sense; and this is intended as a first step in the causal analysis of displays. Thus no attempt is going to be made even to distinguish between the internal and external conditions controlling each tendency, let alone to analyse each still further; moreover, the tendencies are expressed in terms of observed movements.

Ultimately, this first analytical step allows us, as I hope to show, to reduce the problem of the causation of displays to the more general problem of causation of any unitary behaviour system, which in itself seems to me to be important. I also believe that this study of displays is opening our eyes to the fact that dual and multiple motivation is a very common phenomenon—in fact that unitary motivation is extremely rare. This has of course been realised by human psychologists but has received much less attention from students of animal behaviour.

METHODS

The criteria used in our field studies to assess the motivation are of three different, and largely independent, kinds.

(1) Some postures can be seen to be composed of elements of recognisable motor patterns. The Upright offers a good example. As discussed previously (Tinbergen, 79) the raising of the carpal joints, the stretching of the neck and the downward pointing of the bill are characteristic parts of the intention movements of two types of attack: delivering wing beats with the folded wing, and pecking at an opponent from above. Escape components are: coming to a halt instead of continuing to run up towards the opponent (which may occur at different distances from the latter), sleeking of the plumage, upward pointing of the head, and an upward instead of obliquely forward position of the neck. Instead of, or in addition to, coming to a stop, the bird may change its orientation and stand or run more or less parallel to the opponent. In extreme cases a bird may even turn away altogether.

This combination of elements of the attack and the escape patterns can take a great variety of forms. Sometimes attack components are more obvious, sometimes escape components, and many of the Uprights observed can be ranged on a scale from the 'Aggressive Upright' (Fig. 2a) to the 'Intimidated Upright' (Fig. 2b). When an escape tendency is combined with a non-aggressive tendency to stay (e.g. staying near the brood in order to incubate, staying near food) the 'Anxiety Upright' is adopted, in which no signs of aggression can be observed (Fig. 2c).

56

The reliability with which these elements can be labelled with regard to motivation differs from one element to the other. For instance vertical stretching of the neck is also seen as the intention movement of mounting in a male just prior to copulation. The distinction between aggressive and sexual motivation can then be drawn partly on the basis of the other elements combined with it, partly on the basis of 'time scores' as explained below.

Lifting of the carpal joints on the contrary seems to be closely correlated with aggressiveness, not with a general tendency to fly. Arguments in favour of this conclusion are: a bird tending to fly for other reasons than attack (fleeing, flying away to feeding grounds, flying towards the nest in order to relieve the mate) does not lift the carpal joints except at the last moment. A source of confusion is the inertia of this posture: sometimes a bird which, after walking up towards an opponent with carpals raised, is intimidated and turns to flee, will leave the carpal joints outside the supporting feathers. (This is, I believe, because it takes time and an effort to take them in again, and while the opponent is near, this movement is suppressed.) Support for the link between aggression and the lifting of the carpal joints comes from two kinds of comparison: (a) this element is not merely a part of the Upright, but of many other postures which for various reasons must be supposed to contain an element of aggression. Thus it is part of hostile Choking, of the Oblique, of Jabbing, of the Forward, the Mew Call, the Kittiwake's Bow-and-Moan. (b) According to Perdeck's observations on the Great Skua (66a) this species does not raise the carpals in the Aggressive Upright, and this is correlated with the absence of wing beating in fights. This is the more persuasive since the wings are lifted (and spread) during the Oblique-cum-Long-Call.

It seems that there are other elements which are used in more than one posture. Thus downward bending of the head seems to indicate aggressiveness in the Oblique, in the Upright, in Choking, in the Mew Call Posture and in the Bow-and-Moan, while an upwards tendency is often correlated with increasing anxiety or decreasing aggressiveness in the Upright, the Forward, in the Oblique and perhaps in Upward Choking.[1]

[1] Downward pointing of the head need not always be correlated with aggressive tendencies. It is possible that, with increasing precision in the descriptions, we will have to distinguish between different types. (1) Facing Away in the Kittiwake often involves bending down. This seems to be dictated by the position of the other bird from which it turns its face away: if the latter is above the displaying bird, Facing Away is downwards. (2) The posture in which males point out the nest site to females (looking down into the nest cup) and which is often the starting position of 'Friendly Choking', may not indicate aggressiveness, but a

Formal analysis of postures according to this method is of course dependent on a precise knowledge of the motor patterns a species possesses; it is perhaps not superfluous to stress the need for increased accuracy, and the great help which cine films can give; in fact many analyses may not be possible without the aid of filming.

Movements which by this method can be shown to be mosaics of elements of two functional systems have been described in other animals as well (Andrew, 3; Hinde, 30; Marler, 49; Morris, 57).

(2) The second method, that of taking 'time scores', consists of recording which overt behaviour patterns alternate in quick succession with the display studied. The foundation of this method lies in the empirical fact that an animal does not suddenly change from one motivational state into another when the environment is unchanging. For instance, a feeding animal does not suddenly flee in panic unless a sudden stimulus such as the appearance of a predator evokes fleeing (the relatively rare cases of extreme overflow or vacuum activities excepted). In an undisturbed environment a bout of feeding, or a bout of preening, develops gradually (often through a series of intention movements which gradually become more complete) then persists for some time, and then gradually subsides (see, for instance, Van Iersel and Bol, 37). When therefore, in an undisturbed environment, a certain display alternates consistently in a matter of seconds with other movements, we must conclude that the displays and these movements betray roughly the same motivational state. Applied to agonistic displays, this means that we can read their motivation from the fact that they alternate most frequently with complete attack and complete escape or with their intention movements, and not with other movements (except, in many cases, with other displays).

In practice there are some, admittedly quite considerable difficulties. The most important one is that the environment, even with regard to those stimuli that matter, is rarely constant enough. Since these displays act as signals, bird A may change from display-

tendency to look at the brood, and is thus originally a part of the brooding and parental complex (see the discussion of the origin of Choking in the next chapter) (3) Many gulls show the movement of 'staring down'; while this is often followed by pecking at the feet (e.g. in Black-headed Gulls which have stepped on nettles, in many gulls when they have hurt their feet on sharp barnacles) it is also shown in mild conflict situations with an element of fear in it (Goethe, 26). More accurate descriptions will certainly reveal some differences in the exact way the head is pointed down in these different circumstances, and a combination of the method of formal analysis and the two methods to be discussed next will probably allow us to distinguish between these three types of downward pointing with more confidence.

ing to overt attack because bird B has responded to the display, and B's response in turn acts as a signal to A, which then releases an attack. This difficulty can however be overcome. The observer can select for recording those cases where B's behaviour was seen to remain more or less constant. This actually occurs often, for instance when B is strongly aroused to do a third thing, such as, for instance gathering nest material, or trying to force a mating on another bird. A's posturing may then have no observable effect, and B can be taken to be relatively constant. A better method still is to use stationary dummies to release both the display and the attacks or withdrawals; only in this case can one be certain that the relevant aspects of the environment are constant. Even this method can however be applied only when disturbance by outside birds is negligible.

Another difficulty is caused by 'redirected attacks' (Bastock, Morris and Moynihan, 8). Bird A may posture as a response to bird B, then, if B is not obviously intimidated, may shrink from actually attacking it, and attacks a third bird instead. Such redirected attacks must not of course be scored as equivalent to direct attack (although they are interesting in themselves), and therefore cannot be lumped in the time score with direct attacks.

This method, useful particularly in the study of 'displacement activities' (which cannot be judged by method (1)) has already been applied by Moynihan (58), and is now being used more consistently by various other workers. We are beginning to compare the results of observations in the natural situation with those obtained by the use of stuffed birds, and the fact that both methods lead to very similar results shows that the difficulties just discussed can to a certain extent be overcome. Yet the dummy method is of course far to be preferred because it eliminates a highly subjective element: that of deciding (at the moment of observation) whether the situation could be scored as more or less constant or not; and in my experience there are huge differences between otherwise good observers in the ability to do this.

In principle, one can deduce three types of observation from such scores. First, one can see, usually after a few tests, which tendencies are present in the bird and which not. In agonistic displays, the number of complete or incomplete movements recognisable as attacks or withdrawals far outnumbers any other type of movement. In fact, for the distance increasing displays described in chapter IV this evidence is overwhelming, and leaves no room for doubt. Second, the relative proportion of observed attacks and withdrawals linked with each display gives a rough measure of the proportional strengths of the two tendencies. Such scores can be obtained in two ways. As a first, relatively crude (but nevertheless very revealing) method one can systematically compare what, in an unbalanced clash, the victorious bird does and what the withdrawing bird does. This already demonstrates clearly, for instance, that the Kittiwake's Choking is relatively more defensive, and its Bow-and-Moan is relatively more aggressive. A more precise method would be to get figures for the proportions between attacks and withdrawals for every time a certain display is observed. Third, some displays are connected with many withdrawals and attacks, while others often appear without a real fight or even intention movements of attack and withdrawal developing. Generally speaking the proportion of occasions in which the display does not lead to open hostilities gives an (inverse) measure of the intensity of arousal of the two tendencies. In other words, these methods can, in principle, give an indication of both the absolute and the proportional or relative intensities of the two tendencies involved.

It must be admitted that the method sounds better on paper than it is when applied. One difficulty is that especially those movements which depend on a very precise balance of conflicting tendencies may not themselves alternate very often with overt movements of either tendency. Before the balance is so far disturbed that 'pure' behaviour appears, another ambivalent movement may occur first. Thus Choking in boundary conflicts between Herring Gulls alternates with Grass Pulling or with the Aggressive Upright much more often than with actual attack. The interpretation of its motivation therefore rests partly on its linking in time with those other displays which themselves can be interpreted by the time-score method, partly on the use of the third method, to be discussed presently.

Moynihan (**58**) made a start with the application of this method in the Black-headed Gull. E. Cullen applied it in the Kittiwake and G. Manley is engaged in refining it (**47**).

(3) The third method compares the external situations which evoke 'pure' attack and 'pure' escape with the situations which evoke threat displays. In territorial species such as gulls this leads to the following consideration. An intruding male evokes 'pure' attack in the owner of the territory (the fact that even in this situation traces of inhibition by fear may be present can for the moment be neglected). But when the owner meets the same intruder on the latter's territory, this induces 'pure' escape. The significant fact is that most threat displays occur on the boundary zones between territories, where the situation is intermediate between these two extremes. Naturally, for this method a knowledge of the situations in which pure attack and pure escape are elicited is indispensable. The method really amounts to

'measuring' the situation after first having 'gauged' the yardstick with respect to both attack and escape themselves.

Since the three arguments are to a large extent independent of each other, it is of great interest that where more than one method is applied they point to the same conclusion. Thus the formal elements of the Aggressive Upright point to an attack-escape conflict; the time scores show that it alternates much more with attack and escape than with any other behaviour; and it is shown most often near the territory's boundaries. Some displays are not themselves made up of elements of attack and escape behaviour; thus Choking and Grass Pulling contain elements of nest building or, in some species, of other non-agonistic behaviour; this phenomenon will be discussed in chapter VI; but it can be said here that methods (2) and (3) show that they too are often motivated by an attack-escape conflict.

These three methods were worked out after a considerable amount of trial and error. In our earlier studies (see Tinbergen, **79**) they were used unconsciously and our awareness of what we were actually doing has only gradually become acute enough to formulate the methods in precise terms. Therefore I cannot present conclusions except in rather general terms.

THE AGONISTIC DISPLAYS

The Upright, the Oblique, the Forward, Choking and Pecking-into-the-Ground are all obviously due to an attack-escape conflict. The Mew Call is more difficult to understand, but its association with attack and escape is certainly much closer than I recognised in 1953. Particularly in the Herring Gull group and in the Kittiwake it is regularly associated with hostile clashes. A confusing circumstance is that (in the Herring Gull group and in the hooded gulls—not in the Kittiwake) it also may precede courtship feeding and feeding of the young; in fact it is often shown in conjunction with any activity near the nest when either female or young are present. But, judging from 'song' (which so often is an expression of hostility and yet attracts females), the fact that females and young can be attracted by the Mew Call is not really relevant to the problem of motivation; it is possible however that the motivation is more complex than that of the other agonistic postures. This touches a general problem which may well have to do with the fact that, as these movements acquired signal function, they have become 'ritualised' and 'emancipated'; this will be discussed in chapter VII, but it must be said here that the motivational analyses are showing that even in such emancipated displays the connection with attack and escape is still demonstrable, and that the form of a display can do more than merely indicate its evolutionary origin: it usually indicates at least parts of the present motivation.

Moynihan has already shown that there are great differences in motivation between the five first-mentioned postures, at least in the Black-headed Gull. Much however seems to apply to the other gulls as well. For instance, the Upright is shown already at low levels of arousal of attack and escape. It is the last of the postures to disappear as the breeding season advances and hostile clashes become rarer. In the meeting ceremony it is the last posture to be seen before the birds relax. As hostility dies down in the course of the morning, the Upright disappears later than the other postures.

Choking, on the other hand, is due to a high level conflict; it is less often seen without overt attack and escape, and it typically appears in prolonged and vigorous fights between neighbours. The same is true of Grass Pulling in the Herring Gull group. Another peculiarity of Choking is that it is strongly stimulated by the sight of the nest or of the prospective nest site. For instance, as long as Black-headed Gulls posture on the communal feeding grounds outside the breeding territories, Choking is extremely rare even in severe clashes, but as soon as the birds settle in the colony it becomes much more common. This is also true of the Herring Gull group. This point will be discussed in another context in chapter VI.

Further, some displays are relatively more aggressive or 'offensive' than others. Thus, in the Herring Gull, Pecking-into-the-Ground (Grass Pulling) is more often linked with attack than Choking; both indicate a conflict between strong tendencies, but the bird which is dominant in the fight will show Pecking-into-the-Ground while the withdrawing bird will turn to Choking. In the Kittiwake, the Bow-and-Moan takes the place of Pecking-into-the-Ground.

Without going into detail I wish to point out that there seems to be a close parallel between the motivation of the various threat displays and their effect on other birds. As mentioned in chapter III, Choking conveys the information that the bird is not likely to attack, but is likely to defend itself if attacked or approached. This is exactly the correct description of its motivational state. Similarly, the Aggressive Upright and Grass Pulling in the Herring Gull convey the information 'He is ready (and even likely) to attack' and that again describes the motivation correctly. Further a Grass-Pulling Herring Gull means business; it is intensely aroused, though at the same time inhibited by fear; if he attacks, he will do so more vigorously than after merely having adopted the Aggressive Upright; and again this is 'understood' by his opponent. Thus we see that both partners in the encounter are nicely adapted to the interindividual function of spacing-out; the posturing bird has conspicuous, unambiguous effectors (the displays) and the other bird has the specific sensitivity, the 'receptory correlate' which gives the postures their usual effects. Of course this is, in principle, nothing new, but what seems new to me is that this correlation is true in such great detail. Nor does this seem to be the whole story. As already indicated in chapter II, one and the 'same' posture can appear in different forms. Thus the Upright of the Herring Gull can be shown with the carpal joints far out, or almost flat against the flanks. The neck may be more forwardly inclined or more vertical; it may be thick or thin. These variations are correlated with variations in the absolute and the relative levels of arousal of the two component tendencies; when the neck is more forwardly inclined, and when it is thicker this points to greater relative aggressiveness. In the Oblique of the Black-headed Gull the neck and head may be more forwardly or more vertically inclined (see Fig. 3e and f); again, this is correlated with more or less aggressiveness as compared with the fear component. These minor variations in the displays are being studied with regard to their motivation by G. Manley; it would be worth checking whether the correlation with function applies even to these details, as I believe it does.

THE APPEASEMENT GESTURES

The Hunched, Head Tossing and Facing Away differ from the threat postures in that aggression is not the major feature in their motivation. The motivation underlying Facing Away and the Hunched seems to be rather similar. In both cases there are many indications of fear, though this is definitely more pronounced in Facing Away. Yet neither of these postures leads to fleeing very often; the bird is usually stationary. It seems that they are the outcome of a conflict between escape and a tendency to approach or to stay put, but the difference with the threat postures is that this latter tendency need not (though it can) be aggression. This has been analysed in Facing Away. This posture is shown regularly by Black-headed Gulls as part of their pair formation ceremony, and the tendency to stay seems to be the outcome of attraction to the opposite sex, often very specifically to one particular bird. Yet it is also shown in hostile encounters, where the tendency counteracting fleeing is aggression. In Kittiwakes, Facing Away is often seen in birds which try to settle on an already occupied nest. Such birds do not show any signs of aggression; they may fight back if attacked it is true, but their main concern is just to sit on the nest. As soon as such an intruder is attacked by the rightful owners, it faces away. Further, young Kittiwakes Face Away when they are attacked by nest mates or by adults, often without showing any signs of aggressiveness themselves. The link between Facing Away and fear is obvious not only from the analysis of the situations in which it occurs, but also from the fact, reported by U. Weidmann (90) of the Common Gull, that it may lead, through intermediates, to a turning away of the entire body, and subsequent walking or flying away. I therefore consider Facing Away to be the intention movement of turning away and fleeing. It does as a rule not develop into full escape because the bird tends at the same time to stay, and this tendency to stay is not specifically and obligatorily linked with one particular major motivation.

The difference with a posture like the Aggressive Upright, or with the element of lifting of the carpal joints which is found in so many threat postures, is worth emphasising. Here the tendency to approach or to stay is specifically aggression, not any tendency involving approach or staying. A situation similar to that underlying Facing Away is found in the 'chin lifting' of many ducks; in the Eider duck, I have observed this in conflicts between fear on the one hand, and either feeding, caring for the young, or sexual attraction to a mate on the other hand—again a tendency to approach or to stay irrespective of the type of overall motivation involved.

It seems that the Hunched is caused in much the same way; the same kinds of arguments apply to it, but our evidence is less complete. Instead of turning the head away, a bird in the Hunched withdraws the neck, and usually the bill remains pointed at the other bird.

Head Tossing is done by all gulls in two different situations. The females and the large young use it when soliciting food from the male or the parents, respectively, and when it stimulates the other bird to regurgitate (as it usually does) it is followed by eating. But it is also used by both sexes as a pre-copulation display, and it must in those cases be sexually motivated. It seems likely that the tendency to mate is then in conflict with another tendency, and fear is the most obvious candidate for this. There are also indications of aggressiveness. When, as is common in the Black-headed Gull, a male is so strongly inclined to mate that he attempts copulation with a strange female, he shows no overt signs of fear nor of aggression, and the Head Tossing is usually left out altogether.

In some North American species, e.g. the Laughing Gull and the Ring-billed Gull, something very similar to Head Tossing is done in hostile situations, often in connection with the Oblique-cum-Long-Call. According to Moynihan's descriptions (60b), this movement is also similar to the 'Throwback' which is part of the Long Call performance of the Herring Gull and the Common Gull; this Moynihan confirmed to me in conversation. It thus seems that the Throwback and the Head Tossing are related; if so, the underlying motivation seems to vary a great deal.

THE PAIR FORMATION CEREMONIES

Pair formation follows much the same general course in all gulls so far studied. It may occur some distance or even far away from the breeding colony, or in the colony itself; either may happen in the same species. In the Kittiwake it usually occurs on the territory itself, which in this species is little more than the extremely narrow ledge on which the nest is built, but in the other species the pair often and even usually forms first, and then the birds select a nesting territory together (although late males may often settle on a nesting territory before acquiring a mate). On the whole, pairs keep together during the breeding season, and while it is still uncertain to what extent paired birds keep together during the winter (in the Herring Gull for instance, mates may separate or stay together, see Drost, **22**; Tinbergen, **79**, **84**), birds tend to mate with the same partner in successive seasons—of course with numerous exceptions, due, as a rule, to the death of one partner.

Pair formation begins by males settling on a temporary (or, in the Kittiwake, permanent) territory. Even in the Herring Gull, where many pairs form on the 'clubs', males show some site attachment to part of the club-ground. On this pairing territory males are on the lookout for passing or approaching birds—males and females—to which they respond by the Oblique-cum-Long-Call (except the Kittiwake, which performs Choking instead). Whereas other males avoid such 'singing' males, unmated females are attracted to them, and alight near them. Both birds then go through a series of (mutual) displays, after which either the male or the female usually flies off. Males fly round, uttering aggressive calls and often attacking other birds in the air, and return soon to their station; females may alight near another singing male or (in later stages of pair formation) return to the same male. The large gulls tend to walk rather than fly on such occasions.

Pair formation display sequences (or 'meeting ceremonies') of four types are shown in Figs. 14 and 15. That given for the Herring Gull differs from that described in my book (**79**). It has become clear that in my earlier observations I have missed the early stages which correspond to those given for the other species, a fact pointed out by Moynihan (**60c**). J. M. Cullen, E. Cullen and I myself could later confirm that the sequence described by Moynihan for the North

Fig. 14. Meeting ceremony of Herring Gull. From top to bottom:
Oblique of male, Mew Call, Choking, Facing Away.

Fig. 15. Meeting ceremonies of Black-headed Gull (left), Little Gull (centre) and Kittiwake (right). (Little Gull after Moynihan, **58**).

67

American Herring Gull is shown by the European Herring Gull and by the Lesser Black-backed Gull as well. Later in the season females often approach males in the Hunched as described by me in 1953.

A glance at Figs 14 and 15 shows most strikingly that the pair formation sequences are made up almost entirely of movements which are either identical or extremely similar to displays known from hostile encounters between males. The gulls thus confirm the courtship theory developed by various authors (Andrew, 3; Baerends *et al.*, 6; Hinde 29, 30, 31, 32; Van Iersel and Tinbergen, unpubl.; Marler, 49; Morris, 53, 55, 57; Moynihan, 58, 59; Tinbergen, 78, 79, 81) for a variety of birds and fishes. Briefly this theory states that courtship originated as, and often still is, the outcome of a conflict between sexual attraction and agonistic tendencies. 'Sexual' may refer to mating motivation, or to an inclination to be near a mate, which with a shorter or longer interval develops into a tendency to mate. The attack and escape tendencies are themselves in conflict; as I have argued elsewhere (Tinbergen, 78, 81, 57) these tendencies, vital for spacing-out, simply cannot but be aroused by the proximity of a fellow member of the same species, even though this may be a potential mate to which the bird is attracted. In species with 'mutual courtship' (Huxley, 35) such as gulls the female's response once she has alighted is very similar to that of the male, but in all gulls it is the females who show relatively more fear than the males.

In the Herring Gull, the Black-headed Gull and the Kittiwake the sequence after alighting consists of one or two threat postures, followed by an appeasement posture. In some species there are two threat postures (e.g. Mew Call and Choking in the Herring Gull; Long Call and Choking in the Kittiwake)—I can offer no explanation of this. A striking difference in the performance of these displays between agonistic encounters and pair formation situations is that in the latter the birds often stand or run parallel to each other, whereas in a hostile clash the two opponents often face each other. This is very striking in a pair of Herring Gulls running parallel uttering the Mew Call, and in the Forward adopted by Black-headed Gulls (see middle). Kittiwakes, while Long Calling, do not point at each other, but point past the partner's head. This is extremely similar to what many species of geese do in the 'triumph ceremony' (Heinroth, 27) when they adopt a forward posture also frequently used in threat, but avoiding pointing the bill at each other. Hinde (32) reports strikingly similar behaviour of finches.

The name 'triumph ceremony' seems to apply to the analogue of what is here called meeting ceremony; in gulls the pair formation ceremony is identical to the meeting ceremony performed by members of a pair at least in the early part of the season; it extinguishes gradually during the breeding season, although in the Black-headed Gull and in the Kittiwake it may be seen in an almost complete form even after the eggs have hatched.

Sideways orientation occurs in agonistic encounters as well; whether in the pair formation sequences fear accounts for it as it does in hostile encounters seems an open question.

The appeasement posture of the Black-headed Gull is Facing Away, which is usually done in the Upright—a more aggressive form in the male, a more timid form, with a thin neck and less downward-pointing bill, in the female. The Herring Gull and its relatives often show Facing Away, though in a less jerky way than the Black-headed Gull, and usually in the Anxiety Upright. At later stages it is also shown in the Hunched and other postures; the whole complex is less rigidly fixed than in the Black-headed Gull. The extreme downward position of the bill in the Black-headed Gull may not be merely the result of an aggressive tendency; it may be the result of ritualisation correlated with the development of the brown mask, for by pointing the bill down even the upper rim of the mask is completely hidden from view. This effect is further safeguarded by raising the white feathers of the nape (see Fig. 13, bottom, bird on left).

The Kittiwake performs a movement which is not seen in other gulls: Upward Choking. Its appeasement function is not well established and neither is its motivation, but I believe that there can be little doubt that it is analogous to Facing Away. The Little Gull, while having the same song as other species (the Long Call) adopts the Vertical (which seems to be the same as the Forward) and then tilts its head away from the partner. The motivation of these displays is not well known. The tilting is reminiscent of a similar movement in various terns, where it occurs in circumstances where one would expect an appeasement gesture (J. M. Cullen, unpubl.).

Thus, in spite of the great difference in detail between the various species (which will be discussed in chapter VII), and in spite of variation within the species (some aspects of which have to do with fluctuations in the strengths of the three contributing tendencies, the general trend is very much the same in all species studied: the song of the male which attracts unmated females is very similar to one of the known agonistic displays; after alighting one or more displays are shown which again resemble agonistic displays closely, and the series is concluded by an appeasement posture. A striking aspect of the pair formation sequence is that the succession of the three or four postures is so rigidly fixed; the various postures have been integrated into one major display.

Concluding this chapter on the motivation of the displays, I may stress once more the tentative nature of our conclusions. It took us a long time to see the various problems clearly and to work out correct and yet workable methods. The work was in some respects handicapped by being done in the field, where adequate experimentation was found to offer unexpected difficulties. Even the time-score method has its drawbacks; while one has to apply it where many birds are present so as to observe as many interactions as possible, the presence of so many birds also caused impurities which considerably reduced the chances of observing undisturbed encounters. On the other hand, the field observations have the advantage of forcing situations upon the birds which may never occur in neatly controlled laboratory conditions, and for the first stage of the work variety of situations is essential for the formation of hypotheses and for the tracing of inconsistent facts—the 'controls' to be compared with the 'natural experiments'.

Yet it becomes clear that amplification by laboratory tests is required; now that the problems are seen a little more clearly such tests are possible and promising. The most suitable experimental method seems to be to *control* rather than to *assess* the unitary behaviour tendencies separately and to create conflicts between them so as to check and work out conclusions based on interpretations of the type offered in this paper. This method has now been applied in our laboratory to various behaviour tendencies in sticklebacks (Tugendhat, **87**); Tugendhat's work goes far to bridge the gap which consisted between the interpretative field work and the laboratory work done by American psychologists (see, e.g., Miller, **51**). In the future, a combination of the two and possibly other methods would seem called for.

VI. The Origin of the Displays

The description and the motivational analysis of the displays have already given us some hints about their evolutionary origin. In general it can be said that our gull studies have so far supported the conclusion accepted by various authors with regard to other species, viz. that displays are 'derived' movements (Huxley, **36**; Lorenz, **43**; Tinbergen, **77**; Hinde and Tinbergen, **34**). While the derivation of some displays is rather obvious, we are still far from knowing the origin of all, and as long as this is so the question of origin deserves continued critical study in every single case.

The origin of some displays can be recognised, even without comparative study, by their similarity to the movements from which they are derived. Thus a study of the Herring Gull alone leaves little doubt that the Aggressive Upright is in origin a mosaic of components of attack and escape behaviour; that Pecking-into-the-Ground is a redirected aggressive movement; and that Grass Pulling contains an element which can be recognized as 'sideways building'.

However, the origin of many displays cannot be recognized in this way, or often no more than a vague suspicion can be arrived at. This appears to be due to the fact, revealed by comparison of many closely related species, that such derived movements have undergone ritualisation, which I will define as adaptive evolutionary change in the direction of increased efficiency as a signal. While the ritualisation of the displays of gulls will be discussed more fully in the next chapter, it is necessary to insert some general remarks here.

The idea of ritualisation is derived from two sets of facts. (1) Comparison of a given display in a number of closely related species allows one to arrange these species according to the degree of similarity of the derived movement to its supposed origin. Thus cranes use a preening movement in their displays; in some species this is very similar to normal preening, in others it is less similar; in the Mandschurian Crane it is only recognisable as preening by comparison with other, in this respect more intermediate species (Lorenz, **43**). Similarly comparison has helped to recognise the inciting movements of female ducks as derived from a dual origin: the intention movement of pecking at another bird, followed by Chin-lifting, which is in many ducks an expression of fear, often with appeasement function (Lorenz, **45**). A third example is the wing preening movements incorporated in the courtship of many male ducks; the Flag Touching of the Mandarin drake can be recognised as preening through less extreme forms shown by other species (Lorenz, **45**).

This alone however would not be convincing, if it could not be shown (2) that in general the less easily recognisable movements were functionally better suited than the others to the task of providing strong, conspicuous and unmistakable stimuli. This conclusion is based not only on a study of the displays, but also on their function and on an analysis of the type of stimuli animals usually respond to. Lorenz (**43**) has pointed out long ago that 'releasers' are so to speak the graphic representations, the crystallisations, of sign stimuli.

These two considerations suggest that divergence and evolution of displays are the result of a secondary adaptation to a new function of derived movements. As such there is a close parallel between them and, for instance, the secondary adaptation of the first pereiopod in many crayfish, originally a locomotor organ mainly, to the new

71

function of catching and crushing prey, to which it has become adapted, for instance, in the Lobster.

So far, two main sources of derived movements have been recognised. (1) Movements immediately evoked by the situation (such as fighting movements evoked by the approach of an intruder into the territory) are performed either incompletely (intention movements) or with another orientation (redirected movements). Such movements are often called 'autochthonous'. (2) Movements belonging to patterns other than those immediately evoked by the situation and which are therefore usually unexpected and functionally out of context or irrelevant. For these reasons such movements, whatever their exact causation, are called collectively displacement activities. Evidence on the gulls in general confirms this, but they also add something to our knowledge.

I will first discuss some elements which are found in more than one posture, and then proceed to deal with the displays. The sequence will be the same as in preceding chapters.

ELEMENTS OF DISPLAYS

Elements common to more than one posture can be classified in fixed components and orientation components.

Fixed Components In at least the Oblique, the Upright and the Forward the position of the bill and head in the vertical plane varies with the relative preponderance of aggressive over avoidance tendencies; a very aggressive bird tends to point the bill down, a very timid bird less so or even points the bill up. The range over which this occurs varies from one species to the other; thus Hartlaub's Gull points the bill up more than the Black-headed Gull; an anxious Black-headed Gull may assume the Anxiety Oblique in which the bill is pointing up higher than usually in the Herring Gull or the Kittiwake.

Another fixed element which varies with the balance of tendencies is the position of the carpal joints. On the whole, in various postures, lifting of the carpal joints increases with aggression–escape conflict of increasing intensity. Again, as we have seen, lifting of the carpal joints can be an element of the Oblique, the Forward, Choking, the Upright and the Mew Call Posture. Similarly, the Kittiwake opens its mouth in all agonistic postures.

Orientation Components Many aggressive postures of gulls show some orientation in the vertical plane. The Oblique of the Kittiwake and the Black-headed Gull is more vertical when it is evoked by a bird flying overhead than when it is on the same level. The Kittiwake's Oblique can even be directed far down when the opponent is lower on the cliff. The Kittiwake's Jabbing is always precisely aimed at the opponent. In the Oblique therefore the position in the vertical plane is controlled both by the attack–escape balance and by orientation.

Another source of variation is the orientation in the horizontal plane. When aggression is predominant the bill or the body axis tend to be pointed at the opponent, but when a tendency to move away from another bird (caused either by fear or by other tendencies) increases, the bill, the head and neck, or even the whole body turns away sideways. This is clear for instance in the Upright of the Herring Gull: with decreasing predominance of aggression, a bird tends to turn off sideways, and such a bird often walks up to another in a spiral, or even may walk round it in a circle. The Head Flagging movement involves only the head in many postures of the Black-headed Gull; in the Common Gull it may involve the head alone, or the entire body, and Weidmann (90) reports that all intermediates can be observed. The Forward of the Black-headed Gull often shows sideways orientation of the whole body; this is particularly clear in the pair formation sequence (Figs. 13, middle; 14). Choking birds often turn their sides to each other, and in the Black-headed Gull it is not at all uncommon to see two males turning round entirely until they face away from each other.

The effect of a growing avoidance tendency on the head is particularly interesting. While aggression makes a bird point the bill towards the opponent, preponderance of fear and perhaps other avoidance tendencies may result in four different movements which are all the opposite of the intention movement of pecking. (a) The head may be turned in the horizontal plane (Facing Away). (b) The neck may be withdrawn, while the bill still points towards the opponent (Hunched). (c) The bill may point upward (Bill-up Oblique Bill-up Forward, Bill-up Upright, Upwards Choking). (d) The head can be bent down through the vertical until the bill points down and away from the opponent (Facing Away or Beak-Hiding of the Kittiwake).

These components may appear as parts of displays in their pure, non-ritualised form, or they may themselves have undergone ritualisation. For instance, whole Facing Away in the Kittiwake is beautifully oriented every time a Kittiwake shows it, in the pair formation displays of the Black-headed Gull the movement is fixed to the extent of having no orientation in the vertical plane, although in the horizontal plane the orientation is near perfect. As a result, no oriented downward movement of the head can be observed. It is possible that the head-down phase of the Herring Gull's Long Call is originally an aggressive movement; at the moment it does not seem to obey exactly the varying intensities of aggression, but is an integral part of the full Long Call sequence.

THE FULL DISPLAYS

The Upright The analysis of the form of the Upright shows its origin more clearly than that of any other posture. It must have originated as a mosaic of intention movements of attack, inhibited by escape tendencies. The two attack components recognised are (1) stretching the neck and pointing the bill down as a preparation to a downward peck, and (2) the lifting of the carpal joints as the preparation to a wing-beat. The posture seems to be little if at all ritualised.

Two comparative data support this derivation. First, the Kittiwake does not show the Aggressive Upright. Neither however does it have the downward peck as a part of its attack pattern. Second, the Great Skua, while having the Upright, never lifts the carpal joints in the Aggressive Upright. But neither does it fight with wing beating (Perdeck, **66 a, b**). This, together with the fact that in other gulls the lifting of the carpal joints is the specific consequence of aggressive motivation (see chapter v), supports the derivation of this element of the Aggressive Upright.

The Oblique This display, while showing great interspecific variation, is, in all species studied, very dissimilar from any other display. Moynihan, **58**, suggested that it may have been derived from an inhibited forward-attacking movement. I agree that the presence of attack and escape tendencies, as revealed by a study of its motivation, accounts for (a) its orientation—the bill is aimed, at least in the horizontal plane, at the opponent; (b) the frequent lifting of the, carpal joints; and (c) the fact that the Oblique is done while stationary, less often during an actual advance—if during an advance (as when it is adopted in the air) the attack is rarely driven home. All these elements however are present and in fact make up the Jabbing posture of the Kittiwake, the equivalent of which can be found in other gulls as well. But the Oblique has in addition features such as the stretching of the neck and the wide opening of the bill. These, I suggest, are the mechanical corollaries of the loud and long-drawn Long Call. It would thus seem that the Oblique is an aggressive advancing movement inhibited by fear, combined with and modified by the Long Call.

As described in chapter III, the Oblique is, in many species, really a sequence of different postures. After an initial obliquely upward pointing phase (present in most though not all species) the head is bent down (most clearly in the Herring Gull group) then thrown up (either once as in the Herring Gull and—more emphatically—in the Common Gull, or repeatedly as in the Ring-billed Gull), after which a series of long calls is given during which the head gradually comes down to the obliquely forward position. Since, in general, pointing the head down reveals in most gulls a preponderant aggressive motivation, and head-up components a tendency to withdraw, it is possible that the phases of the polyphasic Obliques may have taken their origin in corresponding slight changes in motivation during the performance, and may then have been exaggerated during ritualisation (which in this posture seems to have promoted inter-specific divergence, perhaps because it serves as song, see chapter VII).

Of particular interest is the Throwback phase, since in some species it is remarkably similar to Head Tossing. Such species as Franklin's Gull suggest that the Throwback and Head Tossing have the same origin, and that their dissimilarity in the Herring Gull is the consequence of later divergence between the song, and the pre-copulatory display.

The Forward Up till 1957 I assumed that the Forward of the Black-headed Gull and related species was homologous with the Jabbing of the Kittiwake, and was a slightly ritualised version of the intention movement of a forward-aimed peck. Moynihan (**60b**), however, pointed out that the Forward is more similar to the Low Oblique. Certainly the facts that the Forward is accompanied by a muffled form of the Long Call; that in Franklin's Gull and the Laughing Gull the Oblique in a complete Long Call performance grades into a posture much like the Forward; and that the head plumage is not ruffled in the Forward as it is in the Jabbing and in the Gakkering performances of the Ring-billed and Franklin's Gull would indicate that the Forward is more likely to be derived from the Low Oblique.

The Little Gull's Vertical Display (Moynihan, **58**) is probably the homologue of the Forward. Although its form is rather different from the Forward, it is not so different from the Low Oblique from which the Forward seems to be derived. It differs from the Bill-up Upright, with which it has a certain formal similarity, by the extreme vertical position of the neck, and by the fact that, like the Forward of the Black-headed Gull, it is accompanied by a muffled version of the

Long Call. Further it appears in the situations in which the Black-headed Gull assumes the Forward. Finally, the Bill-up Upright is known in the Little Gull, where it is extremely similar to the Bill-up Upright of the other gulls.

Choking The origin of Choking has been the subject of controversy. While Portielje (**67**) and Kirkman (**38**) pointed out that in the Herring Gull and the Black-backed Gull respectively, the movement was rather similar to that of regurgitation, I argued (**79**) that in the Herring Gull it is in several respects more similar to the movement of depositing nest material. Since it appeared to be causally and functionally out of context (it is caused by a conflict between aggression and fear, and it is not a fighting movement but one belonging to another functional pattern), I have called it displacement depositing of nest material. Since then, however, Manley (pers. comm.) has observed the Black-headed Gull more closely, and has seen that at least in some situations (particularly when a male Chokes at the nest site in the presence of a female) regurgitation is a common part of Choking, and he considers that Choking must be derived from the regurgitation movement.

Comparative data so far known seem to suggest the following interpretation.

Choking, in all species, begins with bending down and pointing the bill down. Further, it is always stimulated by the nest site, as described in chapter v. Thirdly, the initial bending-down movement is taxonomically much more widespread than the full Choking movement: neither skuas (Perdeck, **66a**), nor terns (Tinbergen, **75**; Baggerman *et al.*, **7**; Bergman, **10**; J. M. Cullen, unpubl.) nor most waders (Von Frisch, unpubl.) show anything resembling rhythmic Choking, but many if not all of them stand in a bent posture over the nest or scrape in situations similar to that in which a pair of gulls perform Choking (though not in boundary conflicts). Finally, while in the Kittiwake Choking closely resembles depositing of nest material, in other gulls the rhythmic repetition is much more pronounced in Choking than in depositing, and in the Black-headed Gull Choking is often followed by regurgitation. Further, the Black-headed Gull often alternates the rhythmic Choking call with the Mew or Crooning Call.

In all species where Choking occurs, except in the Kittiwake, the rhythmic repetition of the down–up movements of the head is much more prolonged in Choking than in either depositing or regurgitation, though both do often involve some such movements. This prolonged repetition must be the result of ritualisation; rhythmicity is a common aspect of derived releasers (Morris, **56**).

76

All these facts combined render it likely that Choking is really a combination of two movements. The primary, taxonomically widespread movement is bending down over the nest site. This is very similar to what all these birds do prior to settling down on the eggs and to turning them. It therefore is most probably an incubation movement, which (like all incubation movements, Beer, unpubl.) appears well before the eggs are laid. This is supported by the fact that the posture is facilitated by the presence of the nest or scrape.

Once in this position, a bird which finds itself in a conflict situation (as it is when in the presence of either a female or an opponent) is likely to perform any movement which is facilitated by this initial movement. It is becoming clear from various studies that this type of facilitation (whether by external stimuli, or by proprioceptive stimuli, or by 'central nervous commands' has not been analysed) may dictate which movement shall follow (J. M. Cullen, unpubl.; Lind, 42). Both depositing and regurgitation begin with a head-down posture, and thus might be equally ready to appear. It is possible to decide which of the two is the most likely origin of the rhythmic second phase of the Choking?

The fact that in the Kittiwake Choking resembles depositing so closely is due to the fact that depositing in this species involves very long series of down–up head movements. But this is doubtlessly due to the stickiness of the nest material, while the prolonged rhythmic Choking, found in all species, must be considered an aspect of ritualisation. Thus the similarity, in the Kittiwake, between depositing and Choking has probably developed under the influence of two different kinds of selection pressure and so must be discounted as an argument relevant to the origin of Choking.

J. M. Cullen, 20 has reported that the Black and Brown Noddy (*Anous stolidus* and *A. tenuirostris*) make a movement resembling Choking. The fact that these species are the only known species of terns which build nests and carry material to it, and which regurgitate, confirms the derivation from one of the two movements, but does not allow a decision between them.

The Great Skua does not have Choking (Perdeck, 66a). Neither does it carry material to its nest (it merely scrapes and builds sideways), but it does feed the female and the young by regurgitation. This points towards depositing as the origin of Choking, and pleads against regurgitation.

It seems most likely therefore that the rhythmical movements of Choking have been derived either from depositing or from regurgitation, or even from both, since either or both are facilitated by the initial bent posture. It is quite possible that it is in principle incorrect to say that it is derived from *either* one *or* the other, it may rather have been derived from the down–up element common to both. The prolonged rhythmicity must have developed as a further ritualisation of the initial rhythmicity.

As G. Manley (**47**) points out, this view implies that Choking has originated in situations near the nest site. It is also widely used as a threat posture in boundary disputes. While it is true, and consistent with our view, that even in hostile encounters it is done almost exclusively on the territory and not or far less on communal grounds, hostile Choking is often done many yards away from the nest. Here the initial posture may have to do with the general tendency to bend the neck and head down which normally appears in aggressively motivated gulls, and this posture (although taken up for different reasons than at the nest site in the presence of a female) may facilitate the rhythmic component just as looking into the nest does. This suggests that hostile Choking has developed through 'friendly' Choking.

In some species hostile Choking and friendly Choking show some interesting differences. Hostile Choking in the Black-headed Gull is often done with the wings spread (Fig. 7); this occurs just after a sudden dash at an intruder. This is observed only very rarely in friendly Choking, and probably only in males which show a high level of aggressiveness in other ways as well. Neither does hostile Choking, as a rule, develop into regurgitation. Noble and Wurm (**62**) report that the Laughing Gull ruffles its plumage during hostile Choking which it does not do in friendly Choking. These are elements which probably have to do with the different situations in which the two postures occur, and thus with differences in motivation; in friendly Choking sexual tendencies may contribute which are absent in hostile Choking.

Upward Choking   While it seems plausible to assume that Upward Choking has been enhanced by selection favouring an appeasement movement clearly different from the preceding postures in the meeting ceremony, the origin of the upward posture of the head is obscure. As I pointed out, there is a tendency to point the bill either up, sideways, or deep down, or a tendency to withdraw the neck, in all postures where a tendency to move away is present, or at least a tendency not to point the bill at another bird (which is an aggressive component). This may have made the Kittiwake point its bill up at the end of a meeting ceremony, and the Choking might thus continue with the bill pointing more and more up. At the same time, the bill is closed; this correlates with the waning preponderance of the aggressive tendency, which latter tendency in the Kittiwake is always expressed by opening the mouth.

Another possibility is that the Upward Choking is derived from a swallowing movement; this would be in line with the idea that Choking itself is originally regurgitation. However, normal swallowing does not usually involve an upward movement of the head. It seems to me most probable therefore that the upward tendency is essentially the same as that in several other postures, and indicates a growing preponderance of non-aggressive motivation.

The Mew Call   The origin of this call and posture is unknown.

Pecking-into-the-Ground   In 1953 I concluded, from the obvious similarity of the Grass Pulling of Herring Gulls to collecting of nest material, that it was 'displacement collecting of nest material'. I pointed out, however, that the movement was combined with actual attack movements: the pecking was as vigorous as in actual attack and much more vigorous than ever seen in actual collecting; and if the bird got hold of a firmly rooted plant, it would pull at it as hard as it pulled at an opponent once it has got a grip on its bill, tail, or wing. I described that by writing that actual attack movements were superimposed on the displacement activity. A. Kortlandt pointed out to me that one could just as well and perhaps even better reverse this statement, and say that the displacement activity is superimposed upon attack movements. Our later observations on other gulls suggest that Kortlandt's view is correct. Many species of gulls give vigorous pecks at the ground in hostile and pair formation situations, and in the latter such pecks occur when the bird shows other signs of aggressiveness. Even in the Herring Gull group Grass Pulling usually begins with a vigorous peck. But the Black-headed Gull never or very rarely handles the nest material after such a peck; it fairly regularly performs bill wiping, extremely similar to the movement by which it cleans the bill after feeding in muddy soil or after regurgitation when dirt or food sticks to the bill.

This points to a similar interpretation as that suggested for Choking: Grass Pulling is a dual movement, made up of an initial component (pecking down) which has a wide taxonomic distribution, and a subsequent displacement activity which is different in different species, but which in each case is facilitated by the initial posture. The initial movement is clearly recognisable as an attack movement. It is however 'redirected' (Bastock, Morris and Moynihan, 8; Moynihan, 57): instead of attacking the opponent or partner, the bird attacks the ground. This may seem rather odd, but it is less so when one remembers (a) that it is only little more extreme than the redirected attacks aimed at birds which themselves did not arouse the bird's aggressiveness and which were tolerated until another bird provoked (and yet inhibited) aggression, and (b) that we ourselves, when under the control of irrational anger and simultaneous inhibition (by fear, or by social pressure) do something very similar; we bang the table with our fist, or kick a chair. Allowing for the different motor patterns of attack in man and gulls, I submit that the deeper causation (inhibited irrational anger) is the same in both cases. In Pecking-into-the-Ground therefore we have an example of an 'autochthonous', but complete and redirected movement, followed by a movement originally belonging to another system (nest building or feeding) which is facilitated by the initial posture.

I cannot suggest an explanation why the Black-headed Gull should follow the initial movement by bill wiping, and the Herring Gull by sideways building, but it may be no accident that this difference runs more or less parallel to that in Choking. Nest building movements seem, for some as yet unknown reason, to be more readily available in the Herring Gull group than in the Black-headed Gull, which is more inclined to regurgitate and to wipe its bill (which is often associated with regurgitation).

In the Kittiwake Pecking-into-the-Ground is usually done on the nest, and it is often followed by picking up nest material. A bird starting to peck while on the top of the cliff may even fly to the nest and take up nest material there.

The Hunched The striking features of the Hunched are its withdrawn neck and the orientation of the bill, which is usually aimed at the other bird. It seems probable that withdrawing the neck is the intention movement of total withdrawal, or alternatively of withdrawing the vulnerable face.

Head Tossing The fact that this occurs in both the large young and the female when soliciting food suggests at least one step towards a solution: in the female, it may be nothing, originally, but an infantile movement which reappears at a later stage of life. Certain observations on human behaviour (see e.g. Prechtl, **68**) have shown that infantile movements under certain conditions occur in adult life, and I suggest therefore that this may have been the origin of the Head Tossing in the female. The Upward component may have to do with a fear tendency and with aiming at the bill of the male (parent), which is the source of the food.

The response of the male to the female's Head Tossing is also interesting: it feeds the female in exactly the same way as it feeds the young. This is another example of a movement moved forward in time, which in this case extends much further back in time than the bending movement which initiates Choking; another interesting point is that in the incubation period, which is between the period of intensive courtship feeding and parental feeding, it practically disappears, though to what extent this is due to lack of stimulation I could not say. The phenomenon of courtship-soliciting and courtship feeding is of course widespread in birds as well as in other animals (see Meisenheimer, **50**; Lack, **39**); in birds it is however less widespread than feeding of the young, which therefore must be considered primary. According to the definition of displacement activities given above, both Head Tossing by the female and courtship feeding must be diagnostically labelled as such; the detailed reasons why such irrelevant movements occur in the various examples given may of course be different in each case and further analysis is certainly needed; but the use of the distinction between intention movements and redirected movements on the one hand and displacement activities on the other hand seems obvious and useful. A further point of interest is that so many displacement activities appear to depend on the primary occurrence of an intention movement which facilitates the performance of the 'displacement activity'.

Facing Away   We have seen that Facing Away signifies a fear component which is in conflict with a tendency to stay which is more general than a specifically aggressive tendency. Weidmann's (**90**) observations on the Common Gull, which show many gradations between mere Facing Away and completely walking off or flying away strengthens the conclusion that Facing Away originated as an intention movement of turning in order to flee.

In conclusion, I would like to stress that it is clear that in the absence of historical documentation the interpretations of the origin of the displays must remain tentative. Yet, because of the interrelations with so many other kinds of evidence, certain hypotheses can be rated as more probable than others on the basis of the degree of consistency with known facts. Extension of these studies over more species and more intensive work on motivation and function will widen the range of facts which can be checked for inconsistencies. It is worth pointing out that work on the evolutionary origin of taxonomically related human languages is faced with very much the same situation and tries to arrive at conclusions in essentially the same way as the comparative ethologist (Thieme, **72**).

VII. Some Evolutionary Aspects
INTERSPECIFIC SIMILARITIES

The facts so far known show that the general behaviour, the organisation of the breeding colony, and the individual displays are very similar in all gulls. Taxonomically these characters considerably expand the total range of characters and thus this similarity strengthens the case for grouping all the gulls together and setting them off from other groups. Phylogenetically the similarities in general behaviour patterns, in the form of the displays, in their causation and function and in their probable origin, together with their wide and often continuous occurrence in the whole group strongly support the conclusion that the gulls are a monophyletic group.

Within the group, the similarity is on the whole greater between the species of sub-groups which have already been recognised as such on the basis of morphology. Thus behaviourally the large gulls form a homogeneous group, from which the hooded gulls and the Kittiwake are each set off sharply; within the large gulls, the Common Gull (and the Ring-billed Gull) are to be separated from the Herring Gull group. Within the hooded gull group, the Black-headed Gull and its closest relatives Hartlaub's Gull, the other Silver Gulls and probably Franklin's and Bonaparte's Gull are a homogeneous set rather different from the other hooded gulls. The present paper will not deal with the taxonomy; a revision is given by Moynihan (**60d**).

If the gulls are a monophyletic group, it must be concluded that the individual displays are each the result of divergent evolution started from a common root, i.e. that they are the 'same' in the sense in which Cain and Harrison (**15**) use this term.[1] They have each descended from either a display already possessed by the common ancestor, or have developed parallel after the ancestral species had broken up.

The more widely occurring, the more continuously distributed, and the more similar a display is throughout the group, the more probable I consider this common origin to be. Thus all the Obliques are undoubtedly of common descent, and so are all forms of Jabbing, of Choking and of Facing Away. It is also very probable for the Upright; but it is less certain for the Mew Call and Posture; and the common origin of the Forward of the Black-headed Gull and the Vertical of the Little Gull is still less certain.

Functional studies, which, together with the establishment of occasional, discontinuous occurrence, would give us a lead to possible convergences, have not so far revealed any clear case of this,

[1] I deliberately avoid the term 'homologous' although of course it is often used in the same sense.

although for instance the Large Hooded Gull (*L. ichthyaëtus*) might well be either a Herring Gull which has developed a hood, or a hooded gull which has increased its size, and in either case could be expected to have developed some behavioural convergences.

INTRA- AND INTERSPECIFIC DIFFERENCES

As described in chapter III these are both considerable. Our study of the functional aspects has allowed an assessment of the adaptive significance of these differences. This assessment leads to the conclusion that four classes of selection pressure have been influential in the evolution of the displays: (1) intraspecific differences due to selection pressure on each display separately; (2) intraspecific differences due to selection towards interdisplay distinctness; (3) interspecific differences directly selected for; and (4) interspecific differences developed as a consequence of pressure promoting differences in other functional systems.

(1) As argued in chapter IV there is reason to suppose that certain postures derive some of their peculiarities from pressure favouring conspicuousness as such. Since each posture has developed its own way of achieving this, the increase in conspicuousness has led to intraspecific divergence without selection for distinctness *per se*. Thus the Oblique-cum-Long-Call is directly adapted to the function of giving a loud, long-distance, advertising call, and this has indirectly (so to speak 'accidentally') enhanced its difference from other calls. Choking has developed its prolonged rhythmicity as a means of making the movement more conspicuous and this itself has helped to make it strikingly different from other postures. The Forward may well have become so pronounced and so frequent in the hooded gulls as a correlate to the development of the dark facial mask. The Kittiwake opens its mouth widely in all threat postures as a correlate to the bright orange inside of the mouth; the Great Skua spreads its wings during the Oblique as a correlate to the possession of white wing patches. (By pointing this out, no pronouncement about the evolutionary priority of either *changed* movement or *changed* colour is intended—although occasional spreading of wings, occasional opening of the mouth, and occasional adoption of a 'Low Oblique' is much more widespread taxonomically than the bright colours mentioned, and in this sense the *original* movement is of course older than the bright colour). Alarm calls have developed a staccato quality and rhythmicity.

All these 'changes are to be considered mere improvements of movements functioning as signals; having started from different origins their increased conspicuousness has been achieved in different ways, and thus their distinctness from other signals has been increased as a by-product of selection for conspicuousness alone.

(2) Other differences however must be understood as the outcome of selection pressure favouring differences between displays for the sake of their distinctness. This is perhaps most obvious where two sets of postures serve diametrically opposite functions. Thus all distance-increasing postures of a species may as a group show striking differences from all distance-reducing postures. For instance, all threat postures of the Kittiwake involve wide opening of the mouth, but in all friendly postures the mouth is shut, even in those postures (such as Head Tossing) where other species might open the bill. We must assume that the need for displaying the orange mouth in hostile postures has as a corollary the need for concealing it in non-hostile signalling. A still more specific example seems to be the intraspecific difference which can as a rule be observed between the appeasement postures at the end of the meeting ceremony and the preceding postures. Figs 14 and 15 illustrate the four types of meeting ceremony so far described; in each of them the last gesture is strikingly different from the preceding displays.

In general, the need for unambiguity may well have enhanced the mutual inter-display (intraspecific) distinctness in all displays; one of the means by which this seems to have been achieved is the development of a 'typical intensity' (Morris, 56) which has rendered intermediates between postures relatively rare (see Fig. 1).

(3) The interspecific differences too may have developed either as a consequence of direct pressure towards distinctness, or in a more indirect way. The most likely direct pressure can be expected to have been influential with pair formation displays, for distinctness of these must promote sexual isolation. There is some evidence of this, although the situation is by no means as clear as e.g. in the Anura studied by Blair (12, 13) or the grasshoppers analysed by Perdeck (65), where striking differences in song have been developed. The fact that gulls show far less pronounced differences of this type is no doubt due to the circumstance that indirect pressure has already made some advertising displays of gulls sufficiently different (see below). However, it is possible that some differences may have to be explained as results of direct selection for interspecific distinctness. For instance, the Long Calls of most species are strikingly different. This is even true within the Herring Gull group, where for instance the Long Call of the Western Gull and the Glaucous-winged Gull are different from that of the Herring Gull, whereas all other displays of these three species are identical.

However, the entire problem of sexual isolation among gulls is complicated, since it seems possible that other factors are responsible as well, such as wing-tip patterns (which seem in general to be inter-specifically different within each fauna), size, habitat, and mantle colour; this certainly requires further study. Species-characteristic selection of breeding habitats, which on the whole separates all species, is often overruled by gulls (and terns) being strongly attracted to any gathering of white birds, and occasional visits of individual gulls to colonies of other species are by no means rare: I have myself observed a Herring Gull in an Iceland colony which went through some threat and pair formation displays; in a colony of Hartlaub's Gulls I observed a male Grey-headed Gull (*Cirrhocephala cirrhocephalus*) which displayed to a female Hartlaub's Gull; European Herring Gulls and Lesser Black-backed Gulls often breed in mixed colonies (even with occasional interbreeding); Kittiwakes have been known to breed occasionally in Black-headed Gull colonies.

(4) Some, and perhaps many, differences in displays seem to be the outcome of changes in other behaviour patterns. The best known example of this is the Kittiwake (E. Cullen, 18). Mrs Cullen has shown that many of the Kittiwake's aberrant traits are related to its cliff breeding habit. Her conclusions can be summarised by the list on p. 86, which enumerates and groups the characteristics which she relates to cliff breeding.

The list contains five aspects in which the displays of the Kittiwake are peculiar. First it is the only species which uses the Choking as song, although it has the Oblique. Cullen argues that this may have come about through the fact that the Kittiwake's territory is nothing but the small ledge on which the nest is built. The Kittiwake therefore performs the entire pair formation sequence actually on the nest or the future nest site. We have seen that in all gulls Choking is strongly facilitated by the nest site, and this may originally have favoured Choking as the first threat display to be shown upon the approach of other birds. How this phenotypic effect has been incorporated as an 'innate' (environment-resistant) character is of course an open question.

Now in turn the fact that the first part of the Kittiwake's meeting ceremony consists of Choking, Long Call, and Choking once more may, as we have seen, have been responsible for the specific appeasement posture 'Upward Choking ; the demand for intraspecific distinctness may have made this appeasement gesture so different from that of other gulls. (The fact that the large gulls follow Choking by Facing Away may have to do with the fact that the Choking is not shown as invariably as by Kittiwakes, nor do they necessarily meet very near a nest site.)

Another difference is the absence of the Aggressive Upright in the Kittiwake. This no doubt is due to the fact that the movement from

Breeding habitat: narrow ledges on very steep cliffs = protection from predators

A. Relaxation of other ways of defence:
   1. Tameness
   2. Alarm call rare
   3. Predators not attacked
   4. Chicks not camouflaged
   5. Defecation on nest rim
   6. Egg shells not carried away

B. Protection against falling off cliff:
   1. Strong claws and foot musculature
   2. Female lies down during copulation
   3. Deep nest cup
   4. Two eggs
   5. Immobility of chicks
   6. Chicks face 'wall'
   7. Facing Away in chicks; black neck band

C. Fighting:
   1. No Upright
   2. Special fighting technique ('twisting')

D. Formation of pairs:
   1. Choking as song
   2. (Upward Choking)

E. Nest building:
   1. Collecting of mud; trampling it down
   2. Stealing of nest material
   3. Guarding the empty nest

F. Nest sanitation:
   1. Incomplete regurgitation

G. Parent–infant relationships:
   1. Parents lack food-call
   2. Chicks lack food-'pumping'
   3. Parents do not know chicks individually

which it has been derived, the peck-from-above, is also absent; instead, the Kittiwake pecks straight at the opponent without attempting to bring the bill above him first, and its bill fighting consists mainly of an attempt to throw the other bird off the cliff by twisting.

Thirdly, the alarm call has a greatly heightened threshold, as have all the direct anti-predator responses.

Fourthly, the Kittiwake lacks the 'pumping' movement by which the young of other gulls attract their parents' attention when begging; the Kittiwake has no need of this, since the parent alights on the nest anyway.

A fifth example is the occurrence of Facing Away in Kittiwake chicks, which tends to prevent fights for food between the chicks. Other gulls have not got this appeasement gesture in the chicks; a young when fed simply runs away from the others.

The evidence therefore suggest that, while many displays have undergone specialisation in the direction of their improvement as signals *per se*, resulting in increased conspicuousness, their divergence in evolution seems to be adaptive in three other respects as well: there must have been pressure towards intraspecific unambiguity, pressure directly favouring interspecific unambiguity and indirect pressure through other behaviour, either displays or non-displays.

This analysis of the adaptive character of divergence now allows us to 'strip' species of many of the effects of adaptive radiation and thus, by way of final check, compare the 'cores' of what must be assumed to be older characters. As argued in chapter II, these cores should be more similar than the total complex of characters if the species have descended from a common ancestor. This is actually so in the gulls. When the Kittiwake's Choking is stripped of the adaptive peculiarity (mouth open) it becomes more, and not less similar to that of other species. The Head Tossing, in which no adaptive differences have been discovered so far, is on the whole only very slightly different throughout the group. Once it is realised that the jerkiness of Head Flagging in the Black-headed Gull and its perfect orientation may have to do with the emphatic 'not-showing' the brown mask (which is shown off well by the Forward), and that the presence of Facing Away in Kittiwake chicks is a corollary of its peculiar nesting habitat, this display is much more homogeneous throughout the group than it seemed before a functional appraisal was made. Thus 'peeling off' the obviously adaptive differences strengthens still more the conclusion, already drawn from a mere description of taxonomic characters, that the gulls must be a monophyletic group and that the displays which were given a common name are actually the same.

CONVERGENCES WITH OTHER SPECIES

The preceding remarks are relevant to the use of behaviour

87

characters in taxonomy. Lorenz (**43, 45**) has stressed that displays are often more useful as taxonomic characters than other behaviour. The close similarity of many displays throughout the Laridae corroborates that they can be useful. However, it is easy to overrate this, as the above review of trends promoting evolutionary change emphasises. Considering the relatively small differences in, for instance, feeding behaviour and nest building, one could even argue that the displays have behaved rather *less* conservatively than these patterns. On the average, the truth is probably that displays are neither more nor less conservative than other behaviour patterns.

The recognition of the functional significance of many interspecific differences in displays may also lead to the recognition of the adaptive nature of similarities between distantly related species, and thus warn against the danger of overlooking possible convergences. For instance, Marler (**48**) has shown that the striking similarities in the alarm calls of various small song birds are due to common adaptation to the function of being audible and yet not allowing localisation to the predators of the size which usually prey upon these species. The following illustrates in a slightly different way that similarity of display need not be due to common derivation, unlike (to quote Lorenz's example (**46a**)) the similarities between words of human languages, such as 'father–Vater–vader–padre–père'. Appeasement and related gestures, such as the more 'defensive' threat postures may be very similar in species not at all closely related. In birds which attack with the bill (which itself is doubtlessly a very old character, retained by many different groups), threat postures are often derived from intention movements of this, and involve pointing the bill at the opponent. It is therefore not surprising that postures such as the Forward are found in many birds. Defensive threat postures and appeasement postures, as we have seen, often involve something like doing the opposite of the intention movements of pecking. Now this can in principle be done only in four ways. (1) The head can be turned in the horizontal plane, as in the Facing Away of the Black-headed Gull. (2) It can be turned up in the vertical plane as in the more anxious forms of the Oblique and the Upright, the Upward Choking of the Kittiwake, and the Vertical of the Little Gull. (3) It can be turned down until the bill points backward, as is often seen in the Facing Away of the Kittiwake. (4) Still another way is extreme withdrawal of the bill while it is still pointing at the opponent, and this is typical of the Hunched.

I suggest that it is *because of the limited number of possibilities* of doing 'the opposite' to showing preparedness to attack, that we find some curious similarities in the defensive threat and appeasement postures of widely separated species. Bill-up postures in the extreme form of a 'stretch' are found in such different species as terns, gannets and the Great Tit (Fig. 16). Bending the neck and facing down or even backwards is found not only in Kittiwakes but also in Jackdaws and Coots. Facing Away in a manner more similar than to that of the Black-headed Gull is shown by the Common Crane (Heinroth, 28; Lorenz, 13), and by various ducks (Lorenz, 45). There can be little doubt that these movements, either as defensive threats or as ap-

Fig. 16. Head-up postures in a. Hartlaub's Gull (Anxiety Upright); b. Cape Gannet (2 males in unbalanced clash); c. Great Tit; d. Swift Tern. (a. and d. after Tinbergen & Broekhuysen, 85; c. after Hinde, 29).

peasement postures (they can probably not be sharply separated) occur in many more species (see also Moyniham, 59). Nor need convergences or independent origin of similar displays be rare; it seems quite possible that we are only just beginning to understand their interest and therefore to look for them. Comparison between more widely separated species, with functional studies, would be required to pursue this. It would seem, then, that the chance of convergent similarities in human speech is so much smaller because the number of possible words is so infinitely larger than the number of possible simple appeasement gestures.

EVOLUTIONARY CHANGES IN BEHAVIOUR 'MACHINERY'

Having reviewed and classified the alleged evolutionary changes from the point of view of survival value I will now try and classify them according to the changes in underlying causal organisation which must have been involved. Again, this will be based on the observed intra- and interspecific differences of displays and on the intraspecific differences between displays as derived movements and their presumed origin. Naturally, since so little is known about the behaviour machinery of these displays, this analysis can at best do no more than distinguish between what may (but need not) be different types of change. I believe it is useful to stress the difference between this type of classification and that according to survival value and selection pressure.

(1) The *responsiveness* of many movements and complex patterns *to outside stimuli* must have changed a great deal. This is true for instance of habitat selection: each species selects its own breeding, feeding and roosting habitats. Selective responsiveness to conspecific individuals and categories, to nest material, to food, etc., is also different in each species. The critical distance to evoke territorial hostility is much smaller in the Kittiwake than it is in the Herring Gull, whereas the critical distance in winter is in each species merely proportional to the size of the species.

It is true that, since conditioning occurs so widely in gulls, the innate character of many of these differences is as yet uncertain, but in some instances there are at least indications that we have to do with hereditary changes, as mentioned in chapter II. Yet large-scale exchange of eggs between different species (to study the effect of the environment typical for other species) would be worth undertaking.

(2) Purely *quantitative shifts* in the responsiveness of a movement or a pattern (threshold shifts of some kind) have occurred in both directions. Attacks on predators, escape from predators, and alarm call are either absent or only shown at the strongest provocation by the Kittiwakes, at least as long as they are on the breeding cliffs. The readiness to fly is generally much greater in small than in large species. Young Kittiwakes have strong inhibitions against leaving the nest, and, though less so, against wing flapping. In extreme cases movements have almost or even entirely disappeared, or new movements have appeared. Thus the Kittiwake has lost the aggressive pecking-from-above and with it the Aggressive Upright; but it has 'gained' trampling as part of the nest building (E. Cullen, **18**). In the Great Skua lifting and spreading of the wings has been much facilitated in the Oblique. The same species lacks (i.e. must have lost) the wing-

beating, and consequently the lifting of the carpal joints in aggressive displays.

(3) The *amplitude* of certain movements or parts of movements has often been increased during ritualisation. Thus the Throwback during the Long Call is extreme in the Common Gull; the opening of the mouth in threat postures, especially Choking, has been exaggerated in the Kittiwake. The Black-headed Gull has exaggerated the lifting of the carpal joints; the same species has increased the bending-down in Choking so that it stands in an almost vertical position. Changes of the types (2) and (3) together account for much of the 'exaggeration and simplification' which so many derived movements have undergone during ritualisation.

(4) *Rhythmic repetition* has been evolved or prolonged in several signals: the Long Call, Choking, the Alarm Call. This type of change may of course well be closely related to change in amplitude.

(5) On a higher level of integration elementary displays have often been *combined into one*, either simultaneously or as an internally linked succession. Thus the Black-headed Gull has combined the Upright with Head Flagging (both movements which can occur separately even in this species) into one movement shown at the end of the meeting ceremony. The Ivory Gull has fused what would seem to be the Forward and the Oblique into one rigid series. Several species have added Head Tossing to the Oblique, but while the Laughing Gull Head-Tosses after the full series of Long Calls, the Ring-billed Gull Head-Tosses after each separate call. Pecking-into-the-Ground is successively integrated with Grass Pulling and sideways building in the Herring Gull group, with bill-wiping in the Black-headed Gull. The Bow-and-Moan of the Kittiwake seems to be made up of the Mew Call posture and a bending-down element. The Herring Gull's Long Call has a bending-down component which may well have been added in this way, although it is also conceivable that a bending-down tendency, weak at first, has developed into the extreme bend we observe now.

On a still higher level of integration a long sequences have been formed and more or less fixed; the various pair formation ceremonies are examples.

(6) The *motivation* underlying a posture as a whole may often have changed. This is certainly true, as Morris (**56**) has shown, in all those cases where a 'typical intensity' has been established, i.e. where the range of motivational states in which a display is shown has been widened; this is particularly obvious in displays such as the Forward, the Oblique, and Facing Away, and it may often account for the dis-continuity between displays of the same species, intermediate pos-

tures being rare and often being gone through quickly. It seems possible that many pair formation displays, while originally motivated by sexual attraction mixed with aggression and fear, may have become incorporated more firmly in the sexual pattern and lost part of the original agonistic motivation. This may be so even in those cases where signs of overt aggression and escape are still observable in the meeting ceremony. It also seems that a female approaching a male in the Hunched is motivated sexually mainly, and has very little if any aggressive tendency, nor is fear very obvious in this situation. As reported by Tinbergen and Broekhuysen (**85**) the proportion of aggression and fear underlying the Forward is slightly different in the Black-headed Gull than in Hartlaub's Gull.

(7) The *speed* with which movements are performed can change. Thus the Facing Away in the meeting ceremony of the Black-headed Gull is done much more suddenly than in the other gulls. Head Tossing is repeated much more quickly by the Kittiwake than by the hooded gulls or the Herring Gull group. Extreme slowing down may result in 'freezing': good examples are the 'Soar' of the Black-headed Gull and the Kittiwake, the wing-spreading and lifting of the Great Skua during the Oblique, and the Vertical of the Little Gull.

On the whole, these admittedly tentative and vaguely characterised types of change correspond well with those found in other groups (Blest, **14**; Crane, **17**; Daanje, **21**; Hinde, **32**; Lorenz, **45**; Morris, **56**, **57**). Yet there are some striking differences between gulls and some other groups. Feather raising, so conspicuous in e.g. gallinaceous birds, has not been used to any extent by gulls. Neither have conspicuous colour patterns developed except in a few cases, and gulls cannot compare with such groups as waders, ducks, or gallinaceous birds. It is possible that this is because the white coloration of the ventral parts is of considerable survival value as aggressive camouflage (Craik, **16**; Tinbergen, **79**); this, together with the absence of the need of breeding camouflage (which is found in the eggs and chicks) may also be the reason why colour dimorphism has not been developed. The lack of colourful signal structures may further be the reason why ritualisation has not been so extreme and has not led to such grossly distorted derived movements as in those more colourful groups. Further, although the smaller species indulge in a certain amount of aerial displaying, the larger species miss them almost entirely. Yet the hostile posture and the meeting ceremonies are both elaborate and often repeated. This may have to do with the need of a strong pair bond; it may also be due to the strong competition between the numerous members of a breeding colony. The size of the birds and their colonial nesting, both of which reduce predator pressure, may be the circumstance which have allowed the elaboration of the displays.

VIII. Summary

This paper describes a number of displays of various gulls (Laridae) with special reference to the Herring Gull group, the hooded gulls, and the Kittiwake, and discusses their functions, causation, evolutionary origin and further evolution as signals.

After a sketch of the rationale of comparative behaviour studies (chapter II), the most common single displays and display sequences are described (chapter III). They are rather similar through the family, though many species-specific differences exist. The behaviour similarities strengthen the conclusion that the gulls are a monophyletic group.

Evidence is presented in chapter IV to show that the displays have signal function. A distinction is made between distance-increasing and distance-reducing displays. The reasons for the occurrence of more than one, yet no more than about six distance-increasing postures are discussed; it is argued that each display may well be adapted to deal with a distinct category of opponent: accidental trespassers require another repellent than intentional persistent intruders; and within each category actual trespassers are met in another way than potential trespassers. In addition, one particularly loud call is a typical long-distance threat. This call usually acts as advertisement in a double sense: it repels competitors and attracts unmated females.

Some of the distance-reducing (or appeasement) postures are used in agonistic situations, and still more regularly at the end of the pair formation or meeting ceremony.

The need for and the possibility of more systematic studies of the precise functions of displays is stressed.

Chapter V discusses aims and methods of analysis of the motivation of the displays. The value of 'natural experiments' is stressed. The application of three independent methods leads to the conclusion that agonistic displays are ambivalent, i.e. the outcome of the simultaneous arousal of a tendency to attack and a tendency to flee: the relation between fluctuations in these tendencies and the displays shown is examined. There is a striking correlation between the motivation of these displays and the information they pass on to other individuals.

Appeasement gestures always contain an element of fear; this tendency is in conflict with a tendency to stay, which can be, but usually is not part of the tendency to attack; it may be sexual attraction, attraction to a nest site, or attraction to a provider of food. The need for motivational analyses of the many different forms in which one posture can occur is stressed.

The similarity of motivation in the pair formation ceremonies in the different species is much greater than the formal similarity of the display sequences; the conflict theory of courtship is tested.

The origin of the displays (chapter VI) is varied. Some have clearly arisen as preparatory or intention movements of the patterns directly aroused by the situation ('autochthonous' movements); of these, some are redirected to inanimate objects. Others are derived from movements belonging to functional patterns not directly aroused by the situation ('displacement activities'); their various origins are discussed, and it is shown that they are second components of a dual movement, of which the first component is 'autochthonous' in the above sense and facilitates the displacement activity.

In chapter VII some ultimate causes of evolutionary change are discussed, and a preliminary functional classification of alleged changes in displays is presented. It is argued that change has been enhanced by at least four different types of selection pressure: (1) towards improvement of signal function (conspicuousness); (2) towards increase in intraspecific unambiguity; (3) towards increased interspecific unambiguity; and (4) as a corollary of selection pressure in other functional systems. In some gulls, (4) may have made the major contribution to sexual isolation.

A preliminary classification is given of the postulated evolutionary changes in behaviour mechanisms.

The value of behaviour characters for taxonomic use is considered. After a discussion of adaptive and non-adaptive differences the relative validity of Lorenz's emphasis on the phylogenetic conservatism of displays is reconsidered, and the occurrence of convergent similarities is demonstrated.

REFERENCES

1 ANDREW, R. J. (1956). 'Some remarks on behaviour in conflict situations, with special references to *Emberiza* spec.', *Brit. J. Anim. Behav.*, **4**, 41–5.
2 —— (1956). 'Intention movements of flight in certain passerines, and their use in systematics', *Behaviour*, **10**, 179–204.
3 —— (1957). 'The aggressive and courtship behaviour of certain Emberizinae', *Behaviour*, **10**, 255–308.
4 ASSEM, J. VAN DEN (1954). 'Waarnemingen over het gedrag van de Grote Stern', *De Lev. Nat.*, **57**, 1–9.

COMPARATIVE STUDIES OF THE BEHAVIOUR OF GULLS

5   BAERENDS, G. P. (1950). 'Specialisations in organs and movements with a releasing function', *Symp. Soc. Exp. Biol.*, **4**, 337-60.
6   ——, R. BROUWER and H. TJ. WATERBOLK (1955). 'Ethological studies on *Lebistes reticulatus* (Peters): 1. An analysis of the male courtship pattern', *Behaviour*, **8**, 249-334.
7   BAGGERMAN, B., G. P. BAERENDS, H. S. HEIKENS and J. H. MOOK (1956) 'Observations on the behaviour of the Black Tern, *Chlidonias n. niger* (L), in the breeding area', *Ardea*, **44**, 1-71.
8   BASTOCK, M., D. MORRIS and M. MOYNIHAN (1953). 'Some comments on conflict and thwarting in animals', *Behaviour*, **6**, 66-84.
9   BEER, C. 'Incubation and nest building behaviour of black-headed gulls: I. Incubation behaviour in the incubation period', *Behaviour*, **18**, 62-106.
10 BERGMAN, G. (1946). 'Der Steinwälzer, *Arenaria i. interpres* (L.) in seiner Beziehung zur Umwelt', *Acta Zool. Fenn*, **47**, 1-151.
11 —— (1953). 'Verhalten und Biologie der Raubseeschwalbe (*Hydroprogne tschegrava*)', *Acta Zool. Fenn.*, **77**, 1-46.
12 BLAIR, W. F. (1955). 'Mating call and stage of speciation in the *Microhyla olivacea-M. carolinensis* complex', *Evolution*, **9**, 469-80.
13 —— (1956). '"Call difference" as an isolating mechanism in south-western toads (Genus *Bufo*)', *Texas Jour. Sci.*, **8**, 87-106.
14 BLEST, A. D. (1959). 'The concept of "ritualisation" ', in *Modern Problems of the Behaviour of Man and Animals*, ed. W. H. THORPE and O. L. ZANGWELL, Cambridge.
15 CAIN, A. J. and G. A. HARRISON (1958). 'An analysis of the taxonomist's judgement of affinity', *Proc. Zool. Soc. London*, **131**, 85-98.
16 CRAIK, K. J. W. (1944). 'White plumage of sea-birds', *Nature*, **153**, 288.
17 CRANE, J. (1957). 'Basic patterns of display in Fiddler Crabs (Ocypodidae, Genus *Uca*)', *Zoologica N.Y.*, **42**, 69-82.
18 CULLEN, E. (1957). 'Adaptations in the Kittiwake to cliff-nesting', *Ibis*, **99**, 275-302.
19 CULLEN, J. M. (Unpublished work on the behaviour of the Shag).
20 —— and N. P. ASHMOLE (1963). 'The Black Noddy (*Anous tenuirostris*) on Ascension Island: Pt 2, Behaviour', *Ibis*, **103b**, 423-46.
21 DAANJE, A. (1950). 'On locomotory movements in birds and the intention movements derived from them', *Behaviour*, **3**, 48-99.
22 DROST, R. (1955). 'Neue Beiträge zur Soziologie der Silbermöwe, *Larus a. argentatus*', *Acta XI Congr. Intern. Ornithol.*, 564-9.
23 FRISCH, O. VON. (Unpublished work on the behaviour of waders.)
24 GOETHE, F. (1937). 'Beobachtungen und Untersuchungen zur Biologie der Silbermöwe auf der Vogelinsel Memmertsand', *J. Orn.*, **85**, 1-119.
25 —— (1955). 'Beobachtungen bei der Aufzucht junger Silbermöwen', *Zs. Tierpsychol.*, **13**, 402-33.
26 —— (1957). 'Das Herabstarren, eine Übersprungbewegung bei den Lariden', *Behaviour*, **11**, 310-17.
27 HEINROTH, O. (1911). 'Beiträge zur Biologie, namentlich Ethologie und Psychologie der Anatiden', *Verh. 5. Intern. Ornithol. Kongr. Berlin 1910*, 589-702.
28 —— and M. HEINROTH (1928). *Die Vögel Mitteleuropas*, Berlin.
29 HINDE, R. A. (1952). 'The behaviour of the Great Tit (*Parus major*) and some other related species', *Behaviour Suppl.*, **2**, 1-201.
30 —— (1953). 'The conflict between drives in the courtship and copulation of the Chaffinch', *Behaviour*, **5**, 1-31.

31   —— (1954). 'The courtship and copulation of the Greenfinch', *Behaviour*, **7**, 207–32.

32   —— (1955). 'A comparative study of the courtship of certain finches', *Ibis*, **97**, 706–45; **98**, 1–23.

33   —— (1956). 'Ethological models and the concept of "drive" ', *Brit. Jour. Philos. Sci.*, **6**, 321–31.

34   —— and N. TINBERGEN (1958). 'The comparative study of species-specific behaviour', in *Behaviour and Evolution*, New Haven, 251–68.

·35 HUXLEY, J. S. (1914). 'The courtship habits of the Great Crested Grebe (*Podiceps cristatus*); with an addition to the theory of sexual selection', *Proc. Zool. Soc. London 1914*, 491–562.

36   —— (1923). 'Courtship activities in the Red-throated Diver (*Colymbus stellatus* Pontopp.); together with a discussion of the evolution of courtship in birds', *J. Linn. Soc.*, **35**, 253–92.

37   JERSEL, J. J. A. and A. BOL (1958). 'Preening of two Tern species. A study on displacement activities', *Behaviour*, **13**, 1–88.

38   KIRKMAN, F. B. (1937). *Bird Behaviour*, London and Edinburgh.

39   LACK, D. (1940). 'Courtship feeding in birds', *Auk*, **57**, 169–79.

40   —— (1947). *Darwin's Finches*, Cambridge.

41   —— (1954). *The Natural Regulation of Animal Numbers*, Oxford.

42   LIND, H. (1959). 'The activation of an instinct caused by a "transitional action" ', *Behaviour*, **14**, 123–35.

43   LORENZ, K. (1935). 'Der Kumpan in der Umwelt des Vogels', *J. Orn.*, **83**, 137–213, 289–413.

44   —— (1939). 'Vergleichende Verhaltensforschung', *Verh. deutschen Zool. Gesellsch. 1939*, 69–102.

45   —— (1941). 'Vergleichende Bewegungsstudien an Anatinen', *J. Orn.*, **89**, (Festschr. HEINROTH), 194–294.

46a —— (1953). 'Psychologie und Stammesgeschichte', in G. HEBERER: *Die Evolution der Organismen*, Stuttgart, 131–72.

46b —— (1955). 'Morphology and behavior patterns in closely allied species', *First Conf. Group Proc. Macy Found.*, 168–220.

47   MANLEY, G. (1960). 'The Agonistic Behaviour of the Black-headed gull', D.Phil. Thesis, Oxford.

48   MARLER, P. (1955). 'The characteristics of some animal calls', *Nature*, **176**, 6.

49   —— (1956). 'The behaviour of the Chaffinch', *Behaviour Suppl.*, **5**, 1–184.

50   MEISENHEIMER, J. (1921). *Geschlecht und Geschlechter im Tierreich*, Jena.

51   MILLER, N. E. (1944). 'Experimental studies of conflict', in J. MC V. HUNT: *Personality and the Behavior Disorders*, New York.

52   —— and E. J. MURRAY (1952). 'Displacement and conflict: learnable drive as a basis for the steeper gradient of avoidance than of approach', *J. exp. Psychol.*, **43**, 227–31.

53   MORRIS, D. (1954). 'The reproductive behaviour of the River Bullhead (*Cottus gobio* L.) with special reference to the fanning activity', *Behaviour*, **7**, 1–32.

54   —— (1956). 'The function and causation of courtship ceremonies', in GRASSÉ: *l'Instinct*, Paris, 261–87.

55   —— (1956). 'The feather postures of birds and the problem of the origin of social signals', *Behaviour*, **9**, 75–113.

56   —— (1957). ' "Typical intensity" and its relationship to the problem of ritualisation', *Behaviour*, **11**, 1–12.

57   —— (1958). 'The comparative ethology of Grassfinches (Erythrurae) and

Mannakins (Amadinae)', *Proc. Zool. Soc. London*, **131**, 389–439.

58 MOYNIHAN, M. (1955). 'Some aspects of reproductive behaviour in the Black-headed Gull (*Larus r. ridibundus* L.) and related species', *Behaviour Suppl.*, **4**, 1–201.

59 —— (1955). 'Some remarks on the original sources of displays', *Auk*, **72**, 240–6.

60 —— a (1956). 'Notes on the behavior of some North American gulls I', *Behaviour*, **10**, 126–78; b (1958). II, *Behaviour*, **12**, 95–182; c (1958). III, *Behaviour*, **13**, 113–39;

—— d (1959). 'A revision of the family Laridae', *Amer Mus. Novit.*, no. 1928, 1–24.

61 —— and M. F. HALL (1954). 'Hostile, sexual and other social behavior patterns of the Spice Finch (*Lonchura punctulata*) in captivity', *Behaviour*, **7**, 33–76.

62 NOBLE, G. K. and M. WURM (1943). 'The social behavior of the Laughing Gull', *Ann. N.Y. Acad. Sci.*, **45**, 179–220.

63 PALUDAN, K. (1951). 'Contributions to the breeding biology of *Larus argentatus* and *Larus fuscus*', *Vidensk. Medd. Dansk. Naturh. Foren.*, **114**, 1–128.

64 —— (1955). 'Some behaviour patterns of *Rissa tridactyla*', *Vidensk. Medd. Dansk. Naturh. Foren.*, **117**, 1–21.

65 PERDECK, A. C. (1957). 'The isolating value of specific song patterns in two sibling species of grasshoppers', *Behaviour*, **12**, 1–75.

66a —— (1958). 'Jaarverslag van het Vogeltrekstation over 1957', *Limosa*, **31**, 93–106.

66b —— (1960). 'Observations on the Reproductive behaviour of the Great Skua or Bonxie, *Stercorarius skua skua* Bruun, in Shetland', *Ardea*, **48**, 111–36.

67 PORTIELJE, A. F. J. (1928). 'Zur Ethologie bzw. Psychologie der Silbermöwe, *Larus a. argentatus* Pont.', *Ardea*, **17**, 122–49.

68 PRECHTL, H. (1950). 'Zur Entstehung von Wackeltics', *Oesterr. Zs. Kinderheilk. und Kinderfürs*, **4**, 362–4.

69 SAUER, F. (1954). 'Die Entwicklung der Lautäusserungen vom Ei ab schall-dicht gehaltener Dorngrasmücken (*Sylvia c. communis* Lath.) im Vergleich mit später isolierten und mit wildlebenden Artgenossen', *Zs. Tierpsychol.*, **11**, 10–93.

70 SCOTT, J. P. and E. FREDERICSON, (1951). 'The causes of fighting in mice and rats', *Physiol. Zool.*, **24**, 273–309.

71 STRONG, R. M. (1941). 'On the habits and behaviour of the Herring Gull, *Larus argentatus* Pont.', *Auk*, **31**, 22–50, 178–200.

72 THIEME, P. (1958). 'The Indo-European language', *Sci. Amer.*, **109**, 63–78.

73 THORPE, W. H. (1958). 'The learning of song patterns by birds, with special reference to the song of the Chaffinch *Fringilla coelebs*', *Ibis*, **100**, 535–71.

74 TINBERGEN, L. (1937). 'Feldbeobachtungen an Zwergmöwen, *Larus minutus* Pall.', *Limosa*, **10**, 12–21.

75 TINBERGEN, N. (1931). 'Zur Paarungsbiologie der Flusseeschwalbe (*Sterna h. hirundo* L.)', *Ardea*, **20**, 1–17.

76 —— (1936). 'Zur Soziologie der Silbermöwe, *Larus a. argentatus* Pont.', *Beitr. Fortpfl. Biol. Vögel*, **12**, 89–96.

77 —— (1952) ' "Derived" activities; their causation, biological significance, origin and emancipation during evolution', *Quart. Rev. Biol.*, **27**, 1–32.

78 —— (1952). 'A note on the origin and evolution of threat displays', *Ibis*, **94**, 160–1.

79 —— (1953a). *The Herring Gull's World*, London.

80 —— (1953b). *Social Behaviour in Animals*, London.

81 —— (1954). 'The origin and evolution of courtship and threat display', in *Evolution as a Process*, London, 233–51.

82 —— (1955). 'Psychology and Ethology as supplementary parts of a science of behaviour', *First Conf. Group Proc. Macy Found.*, 75–167.

83 —— (1957). 'The functions of territory', *Bird Study*, **4**, 14–27.

84 —— (1958). *Die Silbermöwe*, Berlin.

85 —— and G. J. BROEKHUYSEN (1954). 'On the threat and courtship behaviour of Hartlaub's Gull (*Hydrocoloeus novae-hollandiae hartlaubi* (Bruch))', *Ostrich*, **25**, 50–61.

86 —— and M. MOYNIHAN (1952). 'Head Flagging in the Black-headed Gull; its function and origin', *Brit. Birds*, **45**, 19–22.

87a TUGENDHAT, B. (1960). 'The normal feeding behaviour of the three-spined stickleback (*Gasterosteus aculeatus*)', *Behaviour*, **15**, 284–318.

87b —— (1960). 'The disturbed feeding behaviour of the three-spined stickleback: I. Electric shock is administered to the food area', *Behaviour*, **16**, 159–87.

88 VEITCH, E. J. and E. S. BOOTH (1954). 'Behavior and life history of the Glaucous-winged Gull', *Walla Walla Coll. Publ. Biol. Sci.*, **12**, 1–39.

89 WEIDMANN, R. (Unpublished work on the behaviour of the Black-headed Gull).

90 WEIDMANN, U. (1955). 'Some reproductive activities of the Common Gull, *Larus canus* L.', *Ardea*, **43**, 85–132.

91 —— and R. WEIDMANN (1958). 'An analysis of the stimulus situation releasing food begging in the black-headed gull', *Anim. Behav.*, **6**, 114.

Section II
Field Experiments

Author's Notes

While all the papers in this section contain a certain amount of description, their common denominator is experimentation in the animal's natural environment. In the earlier work the experiments concerned mainly relatively simple questions of causation, with the emphasis—natural in field studies—on external stimuli that either elicit or orient behaviour. In the later papers another type of experiment was added: one that concentrates on the *effects* of behaviour, and their influence on its success. This shift reflects my growing conviction that such 'survival value' studies, if they are to outgrow the status of speculation about correlations, require experimentation just as much as cause–effect studies aiming at unravelling causation.

The first three papers all deal with the responses of the digger wasp *Philanthus* to external stimuli, with a bias towards the problem of homing. There is nothing much original in these papers; the Frenchmen Henri Fabre and Charles Ferton, the American Phillip Rau and the great Dutch naturalist Jac. P. Thijsse had kindled my interest in digger wasps; Karl von Frisch's work had shown me the power of simple experiments in as natural conditions as possible; and Mathilde Hertz's work on the vision of bees—to me still a model of imaginative experimentation—had inspired my first steps. The gist of the two first papers is well known in the English literature, but the third paper contains certain types of evidence of which the relevance is now much clearer than it was then. Its main lesson seems to me that it demonstrates how even simple pavlovian conditioning is selective, the animal itself deciding what it shall learn and what it shall not. The paper also contains an experiment, equally pertinent to learning problems, which shows the vital function of the 'locality study'—an extreme form of exploratory behaviour in the sense of behaviour which creates the opportunity for learning.

The paper on the Grayling butterfly is included for other reasons. It has a wider scope than the *Philanthus* work, and contains a general description of a greater variety of behaviour patterns, with occasional experimental probes, and a little tentative interpretation. Two points are worth emphasising. Firstly, with the recent wide interest among neurophysiologists in the mechanism of 'gating', it seems to me still desirable to provide more straightforward evidence of the switch in sensitivity, of the animal as a whole, from one part of the available sensory input to another, which occurs with a switch in motivation. Evidence comparable to the strikingly different responses to the same set of coloured and grey papers shown by hungry versus sexually motivated males is still extremely meagre; yet the phenomenon must be pretty well universal. The second point of interest seems to me the way in which purely behavioural evidence forces one to see the 'pooling of input' (Seitz's '*Reizsummenregel*') in action, again at the behavioural level. Finally, the admittedly primitive figure symbolising in the simplest possible 'black box' model a phenomenon that is so characteristic of nervous activity has, in my opinion, not lost its significance. This first step seems often to be bypassed in the modern literature, where the term 'analyser' is used so often as shorthand for what is in reality a synthesising mechanism.

The three papers on eggshell removal, however simple and incomplete, seem to me still valuable as a demonstration of both the functional approach to a behavioural problem, and of a step-by-step analysis. They also demonstrate that field work forces one to a multi-pronged attack, and can contribute experimental evidence on survival value; on certain aspects of causation; and on aspects of the ontogeny of a response. The distinction, so often heard even now, between 'analytical studies' and 'whole animal studies' is clearly a confusing, even nonsensical one—perhaps based on a misconception of what really constitutes an experiment. If the presentation to predators and the preyed-upon species alike, of dummies of different colours, or at different distances, and the measurement of their responses, is not 'analytical', then what is? Nor is it really helpful to draw too sharp a line between 'physiological' and other levels of causal analysis. This is in my opinion the type of thoughtless distinction that hampers interdisciplinary collaboration.

Incidentally, it will strike the reader that neither the *Philanthus* papers nor the Grayling paper contain the simple statistical treatment of data found in the eggshell removal papers. When my wasp studies began I consulted first-rate statisticians and found that they could not help me. I am now ashamed (but not much) to admit that this suited me fine. Some inertia, and a delay in absorbing developments in the English literature (with which I am sure my readers will sympathise) made us lag in picking up the new statistics. But the consequences were not as disastrous as might seem at first glance; often the overall consistency of a variety of results strengthens the conclusions reached; for instance the total figures which demonstrate the motivation-dependent differences in responses to colour of the Grayling, while indicative of the wastefulness of our work in terms of man-hours, are so massive as to leave little doubt.

The paper on food-caching by Foxes is admittedly largely descriptive. But it has a rather nice simple experiment at the end; and it also discusses some functional aspects of over-dispersion, or scattering of the Fox's 'caches'. Although in this case its function is obviously the spread of risk against robbery and not, as in so many vulnerable and camouflaged animals, a defence against being found by a predator with a 'set' searching image, it was during this work that we first began to wonder about the peculiarities of the feeding behaviour of many predators that could exert the pressures that force even well-camouflaged prey animals to avoid clumping.

The last paper in this section was our first attempt to tackle this problem in a more systematic way. Its introductory paragraphs try to explain how we patched together a variety of facts (some no more than suggestive indications) in order to formulate a hypothesis about, on the one hand, the part played by 'choice from sample' in the lives of some predators and, on the other, the pressure that this type of predatory behaviour could exert on certain types of prey. Some points in this paper are controversial; methods have later been improved; yet (even though several of my present co-workers thoroughly dislike it) I consider it one of my most fruitful papers, because it forces the reader to think hard about a fascinating field in which Psychology, Ethology and Ecology can meet, and where naturalists have initiated a bridge-building operation. This particular paper illustrates perhaps better than any of the others the value of what one of our most prominent biologists recently called 'creative observation'. I like to call it 'the ability of putting two and two together'. Unsystematic wondering about a number of seemingly disconnected observations, hardly 'substantiated' but unconsciously registered until they fall into a pattern, may not be to the liking of those who respect measurement more than ideas, but, whatever the deficiencies of this particular paper, it has set in motion a number of new studies, and so fills a niche in our science.

(From the Zoological Laboratory of Leiden University)

2
On the Orientation of the Digger Wasp *Philanthus triangulum* Fabr.
I. (1932)[1]

The Mode of Life of *Philanthus triangulum*

A short review of the scattered and fragmentary information on the habits of the 'bee-wolf' (as this species is often called) (Bouvier, **6**; Bouwman, **7**; Bischoff, **3**, **4**; Fabre, **9**; Verhoeff, **28**; Hamm and Richards, **16**; and others) will be useful, not only because this will lead to a better understanding of my experiments but also because I can supply some new information.

In Holland, *Philanthus triangulum* is a true summer animal; the imagines are seen only from the beginning of July to the end of September. Breeding takes place between mid-July and the end of August. My experiments were done at this time of year in the field at a locality not far from Harderwijk, a small town on the Zuiderzee in the Dutch province of Gelderland, where hundreds of wasps had their holes in the sandy floor of an eroded heathland. The first wasps were seen in early July. They are very sensitive to cold and humidity (even more so than the honey bee) and become active only in warm, fine weather. They are rarely to be seen in rainy or even dull weather, when they remain in their holes for the whole day. It is because of unfavourable weather conditions that some of my observations, especially those made in the first few weeks, are somewhat incomplete.

When the imagines hatch they dig their way almost vertically upwards out of the cells in which the pupae have lain dormant all winter, up to seven animals, males as well as females, originating from each nest. The females—with which the following observations are exclusively concerned—did not start to dig nests and collect prey immediately, but roamed about the nesting area for some days. It

---

[1] The second part of the original paper dealt with the fine structure of receptors on the antennae of *Philanthus*, and has not been incorporated in this volume.

seems probable that in this way they became familiar with the visual characteristics of the area.

From the time when they emerged all females revealed their digger wasp nature, by starting to dig incomplete burrows in a variety of places, not by using their 'scraping combs', as has been described for *Bembyx* (e.g. by Bischoff, **4**), but by shovelling the sand away by means of the strong bristles found on the outer side of the tarsal joints (Fig. 17).

These preliminaries did not, however, go as far as real burrow or nest construction. Each period of digging lasted only a few minutes to an hour; then the wasp abandoned the site and began to dig again in a new place. This period of 'obsessive' digging had an erratic, seemingly aimless character. Each wasp behaved in this way for

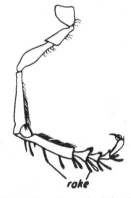

Fig. 17. Right front leg of *Philanthus triangulum*.

several days before it finally completed a proper burrow and started to hunt. One could perhaps interpret this preliminary digging as testing the ground for an acceptable nest site, were it not that it occurred in the same places where, after a few days, numerous real nests appeared. On several occasions I have even observed that a wasp finished a nest previously abandoned by another wasp. I think it is much more likely that we must interpret this behaviour as the 'awakening' of a digging instinct which, in the first days, is not completely integrated with other reproductive actions—a period of gradual functional perfection; a genuine 'juvenile' phase.

I observed the first wasps returning home on July 16th and, shortly afterwards, many wasps were regularly providing for their young. However, the number of working animals kept increasing gradually up to the end of August.

The digging of a new hole took, at the most, about 3 to 6 hours and in no case more than half a day. The hole consisted of a shaft 40–80 cm long, that began almost horizontally but soon ran obliquely downwards, so that the cells or chambers dug at the far end lay about 20–50 cm below the surface. The sand was scooped backwards by the wasp and was left lying outside the entrance, so that each nest was identifiable by an oval patch of sand about 10 cm in diameter. At various places near the end of the shaft, the cells were dug out sideways; once dug, each was filled with 3–6 bees (Bouwman (7) reports 5–8 bees per cell; Verhoeff (28) reports 2). Then an egg was laid on one of the bees and the cell closed off with sand, after which a new cell was dug. After filling a cell (in the course of which the wasp, taking one bee at a time, had flown in and out regularly for some time) a wasp would stay inside for a longer period—which of course slowed down my experiments on homing. After some four or five cells had been completed and provisioned, the wasp abandoned the nest altogether and dug a new one; this was usually close by, but sometimes at a considerable distance, as I established by chance from observations of a marked wasp which, within one week, dug two nests 50 m apart. These observations suggest that the wasps must have to learn the location of each newly dug nest. While the wasp was engaged in the digging of a new tunnel, it came out now and again, flew upwards and hovered over the nest for a few seconds. These short flights were repeated up to ten times before the wasp finally flew off. Then, however, it performed a more elaborate 'departure flight', during which it flew round the nest for a longer period (up to 2 minutes; the average however being about 20 seconds). During this departure flight it turned towards the nest from a variety of directions, flying in wider and wider arcs and rising higher and higher each time, so that it could, and presumably did, scan an ever-increasing area round the burrow.

These elaborate flights on leaving the nest have been described by several observers in a variety of Hymenoptera (Wagner, **3**; Ferton, **10**; the Peckhams, **23**; Opfinger, **22a**). Opfinger, studying honey-bees, was the first to demonstrate their function experimentally. She showed that the term 'orientation flight' (which contains an interpretation of its function) was correct since during it, bees memorise the position of their home in relation to surrounding landmarks. (Training to landmarks in the *immediate* vicinity of the goal occurs, in the honeybee, only during the *approach* flight.)

After the 'departure flight', the wasp went off to its hunting ground where, sooner or later, it caught a bee. The killing of the bee has been described in detail by Fabre (**9**) and my observations agree per-

Fig. 18. Wasp stinging a bee.

fectly with his. For observational purposes, I placed a wasp and a bee under a glass bell (cheese cover). Initially both attempted to fly away and the wasp did not respond to the bee, but soon both became quieter and, when the bee came near enough, the wasp responded, at first merely by showing alertness. But once the bee had (accidentally) touched the wasp's antenna, it at once took up the stinging posture and killed the bee, curling its abdomen in front of the bee's ventral surface and stabbing the bee's head from below (Fig. 18). The bee attempted, in vain, to find a foothold on and to sting the smooth, hard surface of the wasp's abdomen. After a few seconds it was completely defenceless and moved its legs only weakly.

Bischoff (4) describes the hunting method of *Philanthus* as follows: '*Philanthus* stings the captured honeybee very rapidly and waits by its prey after it has fallen to the ground until it has ceased moving.' This is at variance with what I saw.

After having immobilised the bee, the wasp immediately began to rob it of its nectar. It held the head of the victim with its large mandibles and pressed the bee's lower abdomen carefully between two pairs of legs and her own abdomen, and licked up the liquid extruded from the mouth. After repeating this a few times, it took a firm hold of the bee, clasping it beneath her with its head pointing forwards and with the ventral side up, and attempted to fly away with it—in the glass container of course without success.

I was able to observe the homing wasps carrying their prey many times in the field. The wasps never returned with anything but honeybees. This is what most other authors report, but *Halictus* has also been mentioned on occasion; once I found a bumblebee *Bombus terrestris* L. in front of a burrow; this, even if originally taken, had obviously been rejected again by the wasp.

On returning to the nest, the wasp descended slowly from the air and landed almost directly in front of the entrance. This was almost always closed, *Philanthus* usually covering up the nest entrance with sand before departing. The wasp tapped the sand with its antennae and started to dig without letting go of its prey, opened the entrance and, pushing the bee backwards a little, crept inside. After at most 30 minutes, often within 10 minutes, it emerged and, after once more

106

closing the nest opening made a short orientation flight, and flew off to get another bee.

The following incidental observations demonstrate a certain plasticity in a behaviour reputed to be extremely rigid. If a wasp returned with prey to find the nest site in any way disturbed, it sometimes dropped its bee and devoted itself entirely to locating and if necessary restoring the nest. Once this was completed, it walked or flew about, searching for the bee. Once it found it again, it immediately took it up, turned it rapidly into the usual 'transport position' (which required complex manoeuvring) and carried it home. Only once did I observe a wasp sting a bee a second time under such circumstances; in all other cases (at least 25) stinging did not recur.

The wasps dug and hunted regularly up to the beginning of September whenever the weather allowed. Then fewer and fewer wasps came home with bees, although plenty of them were still present on the adjoining heathland. Time and again I was able to observe that on sunny days the wasps, even at this stage, were always active, but were concerned only with digging and did not fly off bee hunting. The digging was similar to that observed at the beginning of July: again incomplete, often interrupted, and now here, then there. They no longer provided for their young; and the wasps remained until their death simply 'digging animals', blindly obeying to the last a fragment of their reproductive drive.

Thus the digging behaviour manifests itself both before and after the season of fully integrated reproduction.

II. Homing and Hunting

The behaviour of the female *Philanthus triangulum* as described above appeared to me to be particularly suitable for the experimental investigation of two problems. The first question was what sensory information the animals used to find their way home with such unerring accuracy, and whether memory played any part in this. Secondly, it was striking that the wasps preyed almost exclusively on honeybees and on no other similar insects. A very great variety of insects were available in the area and consequently the wasps must have been very selective in their choice of prey. These two achievements, homing and recognising the prey, need not of course be controlled in the same way. It is therefore necessary to examine both questions separately.

III. Homing Ability in the Hymenoptera

There are many outstanding works on insect orientation, concerned especially with the homing ability of Hymenoptera. The most detailed are those on the social Hymenoptera, and among these the bees and ants have been particularly well studied. Homing by solitary Hymenoptera has also been studied by many investigators, but since these animals are seldom present in the desired numbers, systematic and complete investigations are lacking. Therefore, before we consider the following experiments on a solitary species, the present knowledge about the social Hymenoptera will be briefly reviewed.

Watson (32) makes a distinction between 'proximate orientation' and 'distant orientation'. He refers to proximate orientation when the animal orients solely in relation to the goal, that is, only with the aid of stimuli coming directly from the goal. As long as it cannot perceive these stimuli, proximate orientation is impossible, and the animal has to resort either to random search or to distant orientation. This classification by Watson possibly corresponds quite well to what occurs under natural conditions. Most cases of distant orientation are attributed to mnemotaxes (Kühn, 18) which, as the name implies, involves memory, and has to involve the use of cues other than those provided by the goal.

Earlier workers repeatedly attempted to explain homing behaviour without reference to memory, but their experiments have not been convincing.

For a long time, there was disagreement even about honeybees (Bethe, 2, versus Buttel-Reepen, 8, and Forel, 15). Recently Wolf (33, 34), using modern techniques, has shown that the bee is capable of relying exclusively on mnemetic orientation, so abolishing the need for Bethe's hypothesis of an 'unknown force', i.e. unknown stimuli emanating from the hive. The ability to distinguish different colours and odours, which Von Frisch (12, 13) has demonstrated in his classic studies, play an important part in the mnemotactic homing of bees.

While in honeybees distant orientation is primarily visual, proximate orientation involves olfactory as well as visual cues; olfactory checking being decisive in the final stages. Wolf devoted many experiments to the 'unknown force', which had been postulated by Bethe, but had already been rejected by, amongst others, Von Buttel-Reepen. The behaviour of bees when their hives are displaced cannot be satisfactorily explained by our present knowledge. Wolf's experiments now indicate that there is an 'ability to register, on the outward journey, the number of turns made' mediated by the antennae, an ability which of course involves memory.

Turning now to the solitary Hymenoptera: as already stated, no large-scale, well-planned and executed experiments have been done and their homing behaviour is therefore still poorly understood.

Fabre (9) describes his observations on *Bembyx*, *Cerceris* and *Chalicodoma*. *Bembyx* immediately finds its nest again if during its absence, alterations have been made to the visual appearance of the nest. In addition, Fabre took a number of *Cerceris* females to a town 3 km from the nest area and released them. Many, but not all, found their way back to the nest. A similar experiment with *Chalicodoma* gave a comparable result. If, on the other hand, a *nest* was displaced over as little as 2 m *Chalicodoma* was unable to find it again and, instead, always returned to the old nest site, following the same trajectory time and again, and searching at the original nest site.

Ferton (10) has published many excellent but little-known observations and experiments on solitary Hymenoptera. In a chapter on orientation and homing ability, he concludes that in *Osmia rufo hirta* and *O. ferruginea*, sight, olfaction and also the capacities ('*les facultés*') of memory and attention ('*l'attention*') play a role in homing. On the basis of Ferton's statement, Piéron (24), however, implies that *O. rufo hirta* uses kinaesthesis ('*sens musculaire*'), instead of Ferton's postulated visual memory for places. I do not consider Ferton's interpretations proven, especially those concerning the perception of odours. Like the majority of researchers in this field he failed to use training methods and relied more on negative than on positive evidence. Yet he describes very clearly how *O. ferruginea* performed an orientation flight over the nest, and suggests that these are made '*afin de s'en fixer dans la mémoire le nouvel emplacement*'.

The Peckhams (23) to whom we owe so many accurate observations on solitary wasps, dedicate a whole section to their 'sense of direction'. They mention and describe 'locality studies' in *Ammophila urnaria*, *Sphex ichneumonea*, *Astata bicolor*, *A. unicolor*, *Cerceris deserta*, *C. nigrescens* and *C. clypeata*. It is obvious from the descriptions that they are concerned here with 'departure flights'. Their conclusion is that the homing of these wasps is guided by their memory of the nest sites.

Bouvier (5) did a training experiment with *Bembyx labiatus* (Fabr.). He laid a flat stone over the nest entrance and, after two days, displaced it about 20 cm. When the wasp returned, it unhesitatingly settled on the 'sham nest'. When, in a control experiment, the stone was replaced in its original position, the wasp selected the correct nest again, thus clearly demonstrating that it had learnt to respond to the stone.

Marchand (21) concludes that *B. rostrata* orients by means of its memory of visual landmarks, but does not provide experimental evidence.

Marchal (20) is even more explicit: he states that, in *Pompilius sericeus*, the learned knowledge of visual cues is sufficient for it to relocate the nest and prey.

According to Wagner (31) the literature contains two diametrically opposed points of view; some investigators believing in a 'sense of direction' and others in a 'memory for places'. Wagner actually takes both explanations to be correct, claiming that his observations on bumblebees provide evidence that the truth lies somewhere between these two extremes.

Finally, we must mention a study by the Raus (26) on *Xylocopa virginica* and *Anthophora abrupta*. According to them these bees are unable to find their nest when they have had no opportunity of learning its whereabouts.

From this short survey, it is obvious that our understanding of the homing of solitary Hymenoptera is far from satisfactory. Experimental investigations are very desirable, and the present paper is intended to provide them at least for *Philanthus*.

IV. Experiments on Homing Ability

A. TRAINING WITH THE 'CONE CIRCLE' STIMULUS COMPLEX

Even simple observations in the field lead one to suspect that the wasps are very sensitive to visual stimuli. If an incautious movement is made within a short distance, say, about 3 m from a wasp which is digging, it flies off. While hunting, a wasp will often fixate a moving bee at some considerable distance, at least 5 cm away. I once observed that a wasp, which had laid its bee down for a moment, obviously noticed it at a distance of some 70 cm, since it flew straight at it and tried to sting it again.

These incidental observations led me to test first of all whether visual stimuli were effective during orientation. A series of experiments was done, inspired by Von Frisch's research on bees (12, 13, 14) in which he used the training method. I tried to make the wasps respond to conspicuous features round their burrows and then, if successful, to find out whether vision was involved.

Experiment 1 For reasons to be discussed in greater detail later, I used pine cones, which lay about on the sandy surface in large numbers and, to all appearances, could well form natural landmarks. I trained the wasps with 20 cones lying in a circle of about 30 cm diameter around the nest (Fig. 19). The cones were laid out between 8 a.m. and 10 a.m. and during the course of the afternoon, the following experiment was carried out. When the wasp had flown

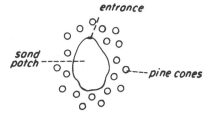

Fig. 19. Experiment 1: training situation.

out to catch itself a bee, I made a sand patch (a 'sham nest') about 30 cm from the real nest, imitating fairly accurately the sandy spot and the slight depression indicating the (covered) entrance. The actual nest was either left intact, or I altered the sandy patch but left the entrance to the nest itself undisturbed. The circle of pine cones was moved so that it surrounded the sham nest (Fig. 20). When the wasp came back and landed on one of the two nests, I drove it off cautious-ly—so that it did not drop the bee—and made it 'choose' again. After I had made at least 5 observations in this way, I immediately con-ducted the control experiment, merely by putting the pine cones back around the real nest. In this situation too, I made the wasp decide at least 5 times. By this control experiment I made sure that nothing but the position of the 'cone circle' had been changed; consequently whatever difference was found in the wasp's choice between experi-ment and control could only be due to the cones' position.

In pilot tests, I had altered other salient features in the vicinity of the nest (the shape and extent of the sand patch etc.) as well as the position of the cone circle because at that time I seriously doubted whether I would be able to train the animals to respond to the cone circle sufficiently well to make them 'blindly' follow this one feature. However, I soon found that the cone circle usually dominated the wasps' choice.

The training was achieved astonishingly quickly; in the 4–6 hours of training, the wasp often flew out no more than 2 or 3 times. I deliberately made the training time so short since, under natural

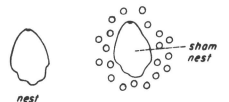

Fig. 20. Experiment 1: test situation.

111

circumstances, a wasp learns to home to a newly dug nest in an incredibly short time; if I had prolonged the training time, the wasp might perhaps orient with respect to cues not normally used.

In comparison with the massive studies of Von Frisch and Wolf, my series of experiments with *Philanthus* are admittedly small. *Philanthus* being solitary, mass experiments on homing were impossible; nor could I lure large numbers to food. For this reason, I could never experiment with more than one wasp at a time, a very time-consuming procedure. Secondly, I never conducted more than one set of trials with any particular individual so as to avoid after-effects that might influence a wasp's behaviour in subsequent experiments. Thirdly, *Philanthus* is very dependent on the weather; it works and collects regularly only on sunny and dry days, of which there were not very many last summer, so that many experiments came to a premature end.

I repeated the first experiment with 17 different wasps and the results obtained can be seen in Table 1. It was demonstrated that all wasps, without a single exception, chose the cone circle on all occasions.

The following protocol provides an illustration. A *Philanthus* nest was provided with a cone circle at 9 a.m., then left unobserved. At 14.25 hrs the wasp returns with a bee, lands on the nest without hesitation and takes the bee in. After 11 minutes it comes out, closes

Table 1. *Results of training with the stimulus complex of the 'cone circle'.*

Wasp No.	Experiment		Control	
	nest	sham nest	nest	sham nest
1	—	9	5	—
2	—	6	5	—
3	—	7	5	—
4	—	5	5	—
5	—	5	6	—
6	—	5	5	—
7	—	7	5	—
8	—	5	5	—
9	—	6	5	—
10	—	8	5	—
11	—	12	5	—
12	—	5	5	—
13	—	5	5	—
14	—	5	5	—
15	—	5	5	—
16	—	5	5	—
17	—	5	5	—
	—	105	86	—

the nest opening and flies off after a barely discernible orientation flight.

I construct a sham nest 30 cm away and place the circle of cones round it.

At appr. 15.00 hrs the wasp comes back with a bee, lands with no apparent hesitation on the sham nest and taps with the antennae on the spot where on the real site there would be an opening under the sand, and it even starts to dig there.

I make it fly up 9 times; every time it selects the sham nest; once it flies about 10 m away, turns into its habitual flight path, but even then selects the sham nest. Then I move the circle of cones back to the real nest, not even attempting to put every cone back exactly in its previous position, but merely trying to replace the cone circle as a whole in roughly the same way as it was during training. Now the wasp selects the real nest, opens the entrance, looks for and finds the bee (which she had left on the sham nest) and flies with it to the nest. Here I make it decide 4 more times, then I let it go in. It leaves the bee outside, crawls into the hole, comes out head first, then grasps the bee in its mandibles and tries to drag it inside, broadside on, which it succeeds in doing after much manœuvring. This incidentally showed once more that there is a greater degree of plasticity in the behaviour of these animals than I had been led to expect.

This experiment therefore shows that the stimulus complex of 'cone circle' was more effective as a landmark than any other feature in the wasp's environment, so much more effective that on no occasion did a wasp 'make a mistake', and I saw only very rarely even a hint of hesitation.

Next I had to determine which of the stimuli from the cone circle were actually used by the wasps. Since their flight seemed to be largely guided by vision the following experiment was done.

B. THE IMPORTANCE OF VISUAL STIMULI IN TRAINING

Experiment 2 During the next phase of training I offered a pine cone circle and *Oleum pini sylvestris*. This oil gives off the smell characteristic of pine cones. A small piece of cardboard, moistened with this oil, was laid on either side of the nest entrance, and a cone circle was laid out as before (Figs 21 and 22).

For the test (Fig. 23), I arranged the cone circle around the sham nest, and placed two similar pieces of cardboard without pine cone scent there, while the papers with the scent remained in exactly the same position. Alternatively, I displaced the scented papers. In

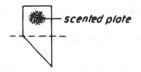

Fig. 21. Scented cardboard plate.

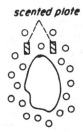

Fig. 22. Experiment 2: training situation.

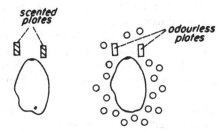

Fig. 23. Experiment 2: test situation.

either procedure the two sets of objects were exchanged as a control.

As before I made each wasp choose at least 5 times in both situations. The scented plates (and especially the non-scented ones) were handled with great care, so that no traces of pine scent could contaminate the odourless plates.

The results of this experiment (Table 2) show that training the wasps with the cone circle succeeded as well as in the first experiment.

Table 2. *Results of training with the cone circle and pine cone scent*

Wasp No.	Experiment		Control	
	nest	sham nest	nest	sham nest
18	—	5	5	—
19	—	5	5	—
20	—	6	5	—
21	—	8	5	—
22	—	5	5	—
	—	29	25	—

114

A weakness of experiment 2 is that the wasps might possibly respond to the particular *concentration* of the scent of real cones and so could have rejected the heavily scented cardboard. To meet this objection, odour had to be eliminated as far as possible, rather than make visual and olfactory stimuli compete.

Experiment 3 Old, weathered cones, which had largely though perhaps not completely lost their scent, were placed in alcohol overnight, then washed in hot water and left to dry in the sun. While the wasps were *trained* with the aid of fresh pine cones, the displacement *test* was done with these, presumably odourless ones.

The results of this experiment did not differ in any way from those of the first two experiments (Table 3). I refrain from giving more detailed data.

Table 3. *Results of training with scentless cones*

Wasp No.	Experiment		Control	
	nest	sham nest	nest	sham nest
23	—	5	5	—
24	—	5	5	—
25	—	5	5	—
26	—	5	5	—
27	—	12	5	—
28	—	5	5	—
	—	37	30	—

Since it was important to eliminate odour altogether as a possible orientation cue, I repeated experiment 1 with antennaless wasps. On *a priori* grounds, as well as from experiments which will be discussed in greater detail later (section v), I assumed that the scent receptors were located in the antennae.

Experiment 4 With the aid of 'bee scissors' (Von Frisch, **14**) and a pair of fine forceps, I amputated, under a magnifying glass, both antennae of wasps which had been trained with the 'cone circle'. Upon release they usually flew off but they soon returned home. In the meantime, I laid out experiment 1 and let these wasps too decide 5 times in both the experimental and the control situations. The results can be seen in Table 4.

Here are some more notes taken at the time of the actual experiment.

One *Philanthus*, trained from about 10.00 hrs to the 'cone circle' complex, returns with a bee at 15.40 hrs. It descends normally and enters its nest without delay, from which I conclude that training has

Table 4. *Results of training antennaless wasps with the cone circle*

Wasp No.	Experiment		Control	
	nest	sham nest	nest	sham nest
29	—	5	5	—
30	—	5	5	—
31	—	5	5	—
32	—	5	5	—
	—	20	20	—

been successful. When it emerges, I catch it and, holding it firmly in the bee scissors, amputate both antennae and release it. It flies away and I quickly arrange the experiment. After a minute it returns. I immediately notice (as Von Frisch, **14**, had done in antennaless bees) that this wasp is flying somewhat unsteadily; almost erratically. Each time it alights I chase it off again; each time it lands and takes off its movements lack the precision of normal wasps; a weak puff of wind for instance is sufficient to throw it off balance. It is allowed the usual 5 plus 5 choices, and each time it chooses the cone circle.

By now it was clear that visual cues near the nest control proximate orientation. It was extremely easy to train wasps to respond to visual landmarks and, from the fact that not a single wasp 'went astray' even once, that is, oriented to other features of the environment, we can conclude that other types of orientation are of negligible importance. Experiments 2, 3, and 4 show that olfactory orientation cannot be very important. Experiment 4 shows, in addition—and this is significant for the evaluation of later experiments—that after amputation of the antennae, the effects of training with visual stimuli remain, so that it cannot be argued that the operation seriously damaged the whole wasp.

In order to check whether other stimuli, apart from visual ones, could be used to guide the wasps to their nests, I attempted to train them to respond to scent alone.

C. TRAINING WITH OLFACTORY STIMULI

Firstly I tried to establish whether the animals responded to an odour at all, and if so, how. After a *Philanthus* had flown out to fetch a bee, I sprinkled the sand immediately in front of the nest entrance with a few drops of pine oil. When a wasp returned with a bee, its behaviour was normal until it came within a few centimetres of its hole. Then it stopped, hesitated a moment, but then landed and even started to dig. But when, while doing this, it tapped on the ground with its antennae, touching the oil-soaked sand, it responded immediately with a twitch of the whole body and a high-pitched buzz, flew up and remained 'hanging' in the air like a hoverfly. I had often seen this on other occasions when flight had been induced, e.g. whenever I moved suddenly. After a short time, it landed again. Often this behaviour was repeated once or twice. Finally it would quieten down, perhaps in part because its digging movements gradually covered the oil patch with fresh sand.

At another nest I dropped oil near the entrance while the wasp was occupied below in its hole. As it emerged, it touched the oilspot with its antennae, and flew up, landed at various other nests, returned to its own nest, was frightened again upon touching the oilspot with its antennae, then, as expected, became used to the odour and, after about an hour, resumed digging and collecting as usual.

It is clear from these observations that the scent was noticed by the wasp, certainly when it touched the oil with its antennae, but perhaps even from a short distance away. Taken together my observations also indicate that the wasps are not guided by any scent coming from the hole itself.

In experiment 1 I was able to establish time after time that when a wasp lands at a sham nest, it starts to dig, repeatedly tapping its antennae on the site of the 'sham entrance'. Such a searching wasp, unable to find the right entrance, often started to dig in a variety of places on the sandy patch, and in the process dropped its bee. Finally, obviously by chance, it might find the real entrance. Since these observations demonstrated nothing positive about the possible role of scent in homing, I did the following experiment.

Experiment 5 Two strips of card scented with pine oil were placed by the nest of a working wasp, one on each side of the nest entrance. After 2 or 3 days, I made a sham nest nearby when the wasp had flown out. Two clean, odourless plates were placed by the real nest and the sham nest received two scented plates. After I had allowed the homing wasp to decide at least 5 times between them, I exchanged the scented and scentless plates and once again allowed it to choose 5 times.

If olfactory orientation were at all possible, we should expect that, after such a long training period with the natural scent of the pine cones, the experiment would give a positive result. As can be seen from Table 5, the attempt to train the wasps to use the scented plates was a complete failure. That some wasps selected the sham nest on a few occasions does not mean very much as the nest and the sham nest were visually very similar. We can also see that, even in the control experiment, a wasp selected the sham nest on one occasion.

Table 5. *Results of training with oleum pini sylvestris*

Wasp No.	Experiment		Control	
	nest	sham nest	nest	sham nest
33	6	1	5	—
34	5	—	5	—
35	7	—	5	—
36	3	2	4	1
	21	3	19	1

The behaviour of these experimental wasps contrasted strikingly with those in the previous experiments. After they had learned to use visual landmarks, I hardly ever observed any hesitation, while wasps in the unsuccessful scent experiments usually showed obvious hesitation. In experiment 5, however, as previously stated, the real nest and the sham nest were visually very similar to one another, while in the previous experiments there was always a considerable visual difference.

One could, however, still raise the objection that the possibility of a scent peculiar to individual nests had not been fully eliminated. Yet this seemed unlikely because:

1. As mentioned before, the wasps often started to dig at the sham nest.
2. They dug, without experimental interference on my part, on more than one occasion at different places at the edge of the sand patch before they found the right entrance of the nest.
3. When, during the wasp's absence, I made drastic alterations to the surroundings of the nest by scraping loose the surface over a circular area of about 50 cm diameter *around* the nest, while leaving the entrance itself intact, the wasp was always completely disoriented, and finally abandoned the nest.

D. FURTHER STUDIES ON THE NATURE OF VISUAL ORIENTATION

1. *Training with Coloured Plates* After I had established that *Philanthus* orients visually when homing, I wished to investigate further the nature of this visual orientation.

As already known, and as I had confirmed by my own observations, the wasps hunt for bees in the latter's feeding areas. They must, therefore, be capable of finding the place where such bees are usually to be found. Hence I was interested in finding out if the wasps, like the bees, are able to recognise the flowers from which their prey collected honey. If they were actually able to do this, one might expect

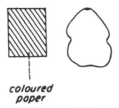

coloured
poper

Fig. 24. Experiment 6: training situation.

them to use the same sensory faculties as bees, both in homing and feeding I therefore wanted to know whether *Philanthus* would use colour when homing.

Experiment 6   Here again I used the training method but instead of pine cones I laid out sheets of coloured Hering papers, each glued behind cleaned photographic plates (9 × 12 cm) so that any possible scent could be washed off. I laid one coloured plate near the entrance to the nest (Fig. 24) and waited until the wasp had become used to it. Then, for the actual experiment I waited for the wasp to fly off, when I constructed a sham nest and laid the coloured plate there, while a plate with one of the graded series of Hering greys was put in the corresponding position near the real nest.

I did this experiment with red (No. 1 of the Hering series), yellow (No. 5) and blue (No. 12) in the same way as Von Frisch (**12**), but obtained only negative results, even after 3 days of training. The wasps alighted almost invariably on the real nest; sometimes a wasp hesitated, but the sham nest was never preferred. I could explain this rather surprising result in one of two ways: (a) either the wasps orient mainly by the shape of landmarks; or (b) they use only 'natural' features, rather than landmarks completely strange to them as Von Frisch had suggested for honeybees.

Favouring the latter explanation I attempted to test this in two different ways.

Experiment 7   A wasp was trained as before, by leaving one plate near the nest for two days. In the test the plate was moved to the sham nest while no plate was put near the real nest.

Experiment 8   As experiment 7, except that the wasp was trained with 2 plates, one one each side of the nest entrance.

For comparison, I trained other wasps with one or two groups of pine cones which covered the same area as the plates (about 9 × 12 cm); the only difference between the paper plates and the 'cone

119

plates' was that the former consisted of flat plates of paper and glass and the latter of pine cones.

Experiment 9   As experiment 6, but with a cone plate instead of a paper plate.

Experiment 10   As experiment 9, but with two cone plates.

Tables 6, 7, 8 and 9 give the data obtained. While perfect training with 'cone plates' was easily achieved, the paper plates produced at best a weak positive response and usually, even after 3 days' training, none at all.

Table 6. *Results of training with one paper plate*

Wasp No.	Experiment		Control	
	nest	sham nest	nest	sham nest
37	5	—	5	—
38	4	3	5	—
39	5	—	5	—
40	4	2	5	—
	18	5	20	—

Table 7. *Results of training with two paper plates*

Wasp No.	Experiment		Control	
	nest	sham nest	nest	sham nest
41	4	1	5	—
42	3	3	5	—
43	5	—	5	—
44	5	—	5	—
45	5	—	5	—
	22	4	25	—

Table 8. *Results of training with one 'cone plate'*

Wasp No.	Experiment		Control	
	nest	sham nest	nest	sham nest
46	—	5	5	—
47	—	6	5	—
48	—	5	5	—
	—	16	15	—

2. *Proximate and Distant Orientation*   We now know, therefore, that *Philanthus* can find its nest using visual landmarks. When, however, it is flying home with its prey, it can see the actual surroundings of the nest from only a short distance, firstly because its

Table 9. *Results of training with two 'cone plates'*

Wasp No.	Experiment		Control	
	nest	sham nest	nest	sham nest
49	—	5	5	—
50	—	5	5	—
51	—	10	5	—
52	—	7	8	—
53	—	5	5	—
54	—	5	5	—
55	—	6	4	—
	—	43	37	—

visual acuity is undoubtedly limited, and secondly because its range is physically limited by the pine trees growing on the plains, where the wasps rarely flew higher than 10 metres.

We are therefore forced to accept Watson's distinction between distant orientation and proximate orientation. In distant orientation, which is extremely difficult to investigate experimentally in *Philanthus*, the wasp finds the area immediately surrounding the nest, where it can switch to proximate orientation. Clearly, my experiments applied only to proximate orientation. The question is: where does distant orientation grade into proximate orientation?

We have seen that the proximate orientation of the wasps can be entirely controlled by a circle of pine cones. If, after training with the cone circles, we move the cones increasing distances away we should reach a stage when the wasp will not be able to find the sham nest, but will either search for or select the real nest.

Unfortunately, owing to unfavourable weather conditions, I was not in a position to undertake a complete series of experiments; but the few results which I was able to obtain would seem to be worth mentioning. They are given in toto in Table 10.

Table 10. *Data from experiment 1 arranged according to distance of landmark displacement. The data from the control experiment are omitted here*

No. of wasps	Distance in cm	Nest	Sham nest
32	50	—	191
2	60	—	10
3	70	—	14
2	100	—	7
1	200	2[1]	—

[1] After a long search.

121

In so far as we can conclude anything from this table, the area in which proximate orientation operates seems to be a circle of a radius of some 1–2 m. Probably, however, this will vary with the physical features and environment of the nest. However, as the following observations show, the area in which proximate orientation is used often contains more than one nest, and this forces the wasp to make a choice.

There are some places where the different holes lie so close together that there are as many as 20 nests in 1 sq. m. More convincing however than this observation is the following. On one occasion I was engaged in several 'cone circle' experiments at the same time. After one of the wasps had flown out, I completed a sham nest with a cone circle about 70 cm west of the real nest (nest 1). There was another nest (nest 2) 1 m east of the real nest which had also been provided with a cone circle, but the training of the wasp belonging to this nest had not been completed. After half an hour, a wasp with a bee arrived and landed twice on nest 2, then again turning into its flight path, began to select the sham nest consistently, and, even in the control, to alight on nest 1. This wasp therefore (which undoubtedly belonged to nest 1) found in its 'nest surroundings' 2 nests surrounded with pine cones. Thus here was an area of at least 200 cm diameter, within which the wasp had to choose; all in all it contained 5 nests.

E. SUMMARY OF EXPERIMENTS ON HOMING ABILITY

This section reports on experiments which show that females of *Philanthus* are able to orient by means of visual landmarks once, through a yet unknown method of 'distant orientation', they have found the 'nest surroundings'. These occupy a roughly circular area of 1–2 m diameter, within which they can be misled by displacement of the landmarks in the immediate vicinity of the nest entrance. Since the surroundings of each nest have to be individually learnt, we are concerned here with an example of visual 'mnemotaxis' (Kühn, 18).

I did not succeed in demonstrating colour vision. It must be emphasised that this in no way implies that *Philanthus* is unable to perceive colour; it is quite possible that, with the aid of 'natural' landmarks, colour vision could be demonstrated.

It could be shown that the wasps orient themselves neither to a specific number of cones, nor to the exact shape of the circle, but to a complex of stimuli, which I have so far not analysed and which for the moment we shall have to call the visual stimulus complex. An attempt to train the wasps to use olfactory stimuli was not successful.

It is perhaps not superfluous to emphasise that these results may not be applicable to other digging wasps. It is, for instance, clear that a homing *Ammophila* behaves quite differently.

My observations cannot decide whether *Philanthus* is able to register and remember the number of turns made on the way out, as Wolf has suggested for honeybees. My only observations on this were that wasps which had been deprived of their antennae appeared to lack the usual fine control of flight movements, as did Wolf's bees which had received the same treatment.

V. Bee Hunting

A. OBSERVATIONS

It is known that *Philanthus* obtains its bees from flowers, that is to say, from places where, in addition to honeybees, many other nectar-feeding insects congregate. It must therefore be able to select its bees from the many other, often similar, insects. In view of what is known of the functioning of the compound eye (see, e.g. Baumgartner, **1**) it seems unlikely that *Philanthus* is able to distinguish honeybees visually from other insects. While I consider it likely that orientation during the search for as well as the actual finding of the prey, is largely a visual process, some other factor must play a part in distinguishing bees from other insects.

Before trying to experiment on this problem I attempted to arrive at a hypothesis by observing the normal, undisturbed bee-catching behaviour.

If, as described in the first section, I put a captive wasp and a bee under a glass bell, the wasp at first did not show any response to the bee but simply tried to escape, flying around aimlessly and banging against the glass. It soon quietened down however, and occasionally settled for a short time. During such rests it might notice the bee, fixating it by following it with movements of the head, and even of the whole body. It did this, however, only for short periods and soon resumed its attempts to escape. But sooner or later the bee and wasp, both flying randomly, would bump into each other. Usually the wasp immediately seized this opportunity to kill the bee. This, I believe, was always triggered off by the bee touching the wasp's antennae. It seemed therefore that stimuli received by the antennae were sufficient to elicit capture even when the wasp was bent on flight. The main stimulus concerned might well be the characteristic odour of the bee by which, I argued, the prey might be distinguished from other insects. I decided to test this experimentally.

B. EXPERIMENTS

The experiments were carried out under the bell. The first aim was to establish that, under these conditions, bees were still preferred to all other insects. I found that the wasps paid no attention at all to various species of flies and bumblebees, but that every honeybee was killed within a short time. This was established with 13 wasps, each of which was given a few flies (mostly Syrphidae) and bumblebees (*Bombus terrestris* L.). The wasp was observed for 1½ hours after a bee had been put into the jar. In all cases, the bees were killed within half an hour, but none of the other insects elicited the stinging response at any time. Since the wasps responded when the bee touched its antennae, I naturally suspected that olfactory stimuli played a part.

Response to smell can be demonstrated in two ways: by denying access to olfactory stimuli (which I did by amputating the wasp's antennae), or by putting bee scent on to other insects which *Philanthus* usually refuses.

1. *Antennaless wasps* One antenna was completely amputated from each of 5 wasps; both antennae were cut off 10 other wasps, while 3 wasps were deprived only of the terminal five segments of each antenna.

For the amputation, the wasps were held in the 'bee scissors'; it was then possible, under a magnifying glass, to cut off pieces of the antennae of the desired size with sharp scissors. On each occasion an experimental wasp was observed for about 1½ hours together with one or two bees. As a control, a normal wasp was always added; if the latter had not killed its bee by the end of the experimental period, a negative result with the experimental animal was discarded.

The 5 wasps from which I amputated one antenna proved to be no slower than the normal wasps and captured and stung all the bees given to them. I observed that, on touching the antennaless side of the wasp's head, the bee evoked no stinging response but caused the wasp to adopt a defensive position: it raised its front legs and opened its strong mandibles. This posture was also adopted by wasps when accidentally bumping into one another.

As I mentioned above, I amputated the last five segments of each antenna in 3 wasps. Of those wasps which had been operated on in this way, two paid no attention to the bees, but the third wasp, which had initially shown a similar lack of interest when touched by a bee, suddenly responded positively: after its antennal stumps had been in contact with the bee for rather longer time than normal it tapped the bee all over quite carefully and attentively and, after about half a minute, showed first a weak, and then a normal stinging reaction, with which it killed the bee in the usual way.

The 10 wasps from which I had completely amputated both antennae were observed for a prolonged period. Not a single one of them ever showed any interest in either bees, flies or bumblebees. All control wasps captured and stung at least one honeybee before the experiment was broken off.

Before drawing conclusions from these data, we must briefly discuss whether the damage caused by the operations could by itself account for the results. The wasps which had been deprived of both antennae lived for at least a week, which is longer than similarly treated honeybees. During this time (as previously mentioned, I carried out such amputations in my work on homing) the animals worked normally, except that they no longer brought any bees home. (My hope that they would come home with prey of different species was not fulfilled.) Either these wasps were unable to carry out the finely controlled movements necessary to catch bees, or the positive stimulus of the bee odour was necessary to trigger the stinging reaction.

In addition, the antennaless wasps used in my homing experiment were, as described earlier, using the pine cones as visual landmarks; they responded just as much to displacements of cones as intact wasps.

I conclude therefore that the absence of response to bees by the amputated wasps was not due to overall damage, or 'shock', and that capturing and stinging can be elicited by stimuli received by sense organs in the antennae. Whether these were olfactory stimuli had to be decided next.

2. *Flies Carrying Bee Odour* While previous studies had shown that the olfactory sense organs of Hymenoptera are indeed located on their antennae, my experiments did not so far provide any proof of this in *Philanthus*. I therefore tested the wasps' responses to bee scent.

First, a number of flies were offered, under the glass dome, to undamaged wasps. During periods varying from 1 to $1\frac{1}{2}$ hours none of these wasps showed any traces of capturing or stinging responses.

I then took the flies out, rubbed them with a bee which I had just killed by crushing it, and returned these 'bee-scented' flies to the done.

I did this with 6 wasps. Of these, 5 responded with the complete stinging sequence, most of them 3 or 4 times in a short period. It was, however, remarkable that not a single wasp actually succeeded in killing its prey; they invariably let the fly go again soon afterwards. It looked as though, at the last moment, the wasp noticed its mistake (perhaps with the aid of a tactile organ on the abdomen), and did not drive the sting home.

The conclusion to be drawn from these experiments is that *Philanthus* is capable of recognising honeybees by means of olfactory stimuli alone. Since under the conditions of the tests the wasps did not respond until actually touched by a bee, and since it seems unlikely that free-roaming wasps would simply wait for this to occur, the full bee-hunting behaviour still remains to be analysed. It is to be expected that visual stimuli have to guide the wasps to their prey before capturing and stinging can occur.

REFERENCES

1 BAUMGÄRTNER, H. (1928). 'Der Formensinn und die Sehschärfe der Bienen', *Z. vergl. Physiol*, 7.
2 BETHE, A. (1902). 'Die Heimkehrfähigkeit der Ameisen und Bienen', *Biol. Zbl.*, 22.
3 BISCHOFF, H. (1923). *Biologie der Hymenopteren*, Berlin.
4 —— (1923). 'Hymenoptera', in P. SCHULZE: *Biologie der Tiere Deutschlands*, Berlin.
5 BOUVIER, E. L. (1900). 'Le retour au nid chez les Hyménoptères prédateurs du genre *Bembex*', *C. r. Soc. Biol. Paris*, 52.
6 —— (1916). 'Quelques observations sur les Philanthes', *Ann. Inst. Pasteur*.
7 BOUWMAN, B. E. (1927). '*Philanthus triangulum* F.', in 'De Graafwespen van Nederland', *De Levende Natuur* 32.
8 BUTTEL-REEPEN, H. VON (1900). 'Sind die Bienen Reflexmaschinen?', *Biol. Zbl.* 20.
9 FABRE, J. H. (1923). *Souvenirs entomologiques*, 80, Ausgabe, 1. und 4. Serie., Paris.
10 FERTON, CH. (1923). *La Vie des Abeilles et des Guêpes, œuvres choisies, groupées et annotées par E. Rabaud et F. Picard*, Paris.
11 FREIDLING, H. H. (1909). 'Duftorgane der weiblichen Schmetterlinge', *Z. Wiss. Zool.*, 92.
12 FRISCH, K. (1914). 'Der Farbensinn und Formensinn der Biene', *Zool. Jb., Abt. Allg. Zool. u. Physiol.*, 35.
13 —— (1919). 'Über den Geruchssinn der Biene und seine blütenbiologische Bedeutung', *Ibid.*, 37.
14 —— (1921). 'Über den Sitz des Geruchssinnes bei Insekten', *Ibid.*, 38.
15 FOREL, A. (1910). *Das Sinnesleben der Insekten. Übersetzt von M. Semon*, München.
16 HAMM, A. H. and O. W. RICHARDS (1930). 'The biology of the British Fossorial Wasps', *Trans. Entomol. Soc. Lond.*
17 HAUSER, G. (1880). 'Physiologische und histologische Untersuchungen über den Geruchssinn der Insekten', *Z. Wiss. Zool.*, 34.
18 KÜHN, A. (1919). *Die Orientierung der Tiere im Raum*, Jena.
19 LEYDIG, FR. (1886). 'Die Hautsinnesorgane der Arthropoden', *Zool. Anz.*, 9.

20 MARCHAL, P. (1900). 'Le retour au nid chez le *Pompilus sericeus*', *C. r. Soc. Biol. Paris*, **52**.

21 MARCHAND, E. (1900). 'Sur le retour au nid de *Bembex rostrata* Fabr.', *Bull. Soc. Sci. nat. Ouest*, **10**.

22 MURR-DANIELCZICK, L. (1930). 'Über den Geruchssinn der Mehlmotten-schlupfwespe *Habrobracon Juglandis* Ashmead', *Z. vergl. Physiol.*, **11**.

22a OPFINGER, E. (1931). 'Über die Orientierung der Biene an der Futterquelle', *Z. vergl. Physiol.*, **15**.

23 PECKHAM, G. W. and E. G. PECKHAM (1898). 'On the Instincts and Habits of the solitary wasps', *Wisconsin Geol. a. Nat. Hist. Survey, Bull.*, **2**, Sci. Ser. **1**.

24 PIÉRON, M. (1906). *Bull. Inst. gén. psychol.*

25 VOM RATH, O. (1888). 'Über die Hautsinnesorgane der Insekten', *Z. Wiss. Zool.*, **46**.

26 RAU, PH. (1928). 'Experimental studies on the homing of carpenter and mining bees', *J. comp. Psychol.*, **7**.

27 RULAND, FR. (1888). 'Beiträge zur Kenntnis der antennalen Sinnesorgane der Insekten', *Z. Wiss. Zool.*, **46**.

28 VERHOEFF, C. (1892). 'Beiträge zur Biologie der Hymenopteren', *Zool. Jb. Abt. System.*, **6**.

29 VOGEL, R. (1923). 'Zur Kenntnis des feineren Baues der Geruchsorgane der Wespen und Bienen', *Z. Wiss. Zool.*, **120**.

30 WACKER, FR. (1925). 'Beiträge zur Kenntnis der antennalen Sinnesorgane der Hymenopteren', *Z. Morph. u. Ökol. Tiere*, **4**.

31 WAGNER, W. (1907). 'Psychobiologische Untersuchungen an Hummeln', *Zoologica*, **19**.

32 WATSON, J. B. and K. S. LASHLEY (1915). *Homing and related activities of Birds*, Publ. 211 from the Carnegie Inst. of Washington.

33 WOLF, E. (1926). 'Das Heimkehrvermögen der Bienen, I. Mitt', *Z. vergl. Physiol.*, **3**.

34 —— (1928). 'Das Heimkehrvermögen der Bienen, II. Mitt', *Ibid.*, **6**.

(*From the Zoological Laboratory of Leiden University*)

3

On the Orientation of the Digger Wasp *Philanthus triangulum* Fabr.:

II. The Hunting Behaviour (1935)

Introduction

In a previous paper reprinted above (**13**), I showed that females of *Philanthus* home to their burrows using visual landmarks. In addition I reported on a few experiments which demonstrated that it can distinguish its special prey, the honeybee, using olfactory cues alone.

In my observation area, this restriction to the honeybee was absolute; in two summers' work in a colony of some 1,500–2,000 wasps I saw approximately 2,000 prey carried home, and these were invariably honeybees. On one occasion I found a small bumblebee worker lying near a nest entrance; presumably it had been captured, but ultimately rejected. Since the hunting areas adjoining my colony were teeming with insects of many species, my wasps must have been extremely selective.

Some authors also mention *Halictus* and *Andrena* spp. as occasional prey; and Molitor (**10**) has published an even more extensive list. By placing a variety of dead insects near the entrances of the burrows he could induce his 'bee wolves' to carry in 'a little bumblebee'; 'a large fly'; 'a *Dasypoda*'; 'a small bee similar to *Anthophora*' and 'a *Dasypode plumigera* female'; yet the only insects actually carried home from the hunting area by the wasps themselves were, as in my own colony, honeybees. Molitor concludes from his evidence 'that digger wasps use the sense of smell to a far lesser extent than ants' (p. 181). While the interesting fact remains that insects which are ignored by *Philanthus* in the field may be accepted when offered near the nest entrance, the observations reported in the present paper show that Molitor's data do not refer to the actual hunting. The complete hunting behaviour turned out to consist of a sequence of activities,

128

each of which is controlled by sensory stimuli specific to it, and this full sequence is shown only in response to honeybees.

My earlier experiments, done under a relatively small glass dome, showed no more than that the last stages of the hunting behaviour, the actual capturing and stinging, could be elicited by olfactory stimuli alone; but these experiments shed no light on how *Philanthus* discovers a prey in the first place. Since the summer of 1934 offered another excellent opportunity to study *Philanthus*, I made a more extensive study of which senses are involved in the full hunting behaviour. To my knowledge, this has so far not been done for any other digger wasp.

My study started through a chance observation of a wasp that had, for some reason or another, dropped her bee after returning to the nest entrance, and retrieved it after finding and opening the burrow entrance. I was surprised to find that, when recovering such an immobilised bee, the wasp responded in a way that was strikingly different from her hunting behaviour towards live prey.

The Recovery of a Dead Bee

Occasionally, e.g. when I made a sudden movement, or when a wasp had difficulty in finding the nest entrance, it would drop its bee. As a rule it would sooner or later recover it, grasp it, quite often sting it again, and carry it home. In such cases a wasp would always approach the bee upwind—sometimes walking, sometimes in flight. Although this suggested orientation by scent, the upwind approach could also have aerodynamic advantages in flight; homing wasps too (which, as I showed previously, use visual landmarks) may cover the last few feet flying low against the wind; on windy days such a final upwind approach may even be the rule.

Experiment 1 A wasp returning home with a bee was frightened off as soon as it had landed on the sandy patch in front of the entrance. This made it drop its bee and fly off. I displaced the bee over various distances at right angles to the direction of the wind and observed the behaviour of the wasp upon her return. In 42 trials, conducted with 23 wasps, the bee was retrieved on 35 occasions. It should not be concluded that on the other occasions the wasp was unable to find its bee, for in every case the wasp switched to digging, and did not respond to the bee even if it accidentally touched it with the antennae; I concluded therefore that the absence of a response was due to a switch in motivation.

These experiments demonstrate, firstly, that not every wasp that has lost its bees tries to retrieve it, and secondly that those that do retrieve it respond to stimuli from the bee itself. This stimulus can be perceived only over a relatively short distance: the usual distance at which a wasp began its upwind course was approximately 1 m; the maximum observed was 2 m. Wasps that failed to approach to within 2 m kept flying round over a wide area centred round the nest; in this they were obviously guided by knowledge of the visual appearance of the general nest area, and they had to blunder into scent trail before they could start orienting chemically.

The following entry gives an idea of the type of observations recorded:

17.7.34. A *Philanthus* homing with a bee is frightened off just after landing on its sand patch, where it drops its bee. I take the bee away and put down a dark pebble of bee size 5 cm from where the bee had been left. The wasp returns immediately to the nest area and flies round in low, wide loops, without selecting any particular place. The circling flights encompass an area of approximately $1\frac{1}{2}$ sq. m and last approximately 30 seconds. I then exchange the pebble for the bee; soon the wasp returns and stops approximately 50 cm downwind of the bee, then proceeds upwind, waving its antennae; it finally pounces upon the bee from about 10 cm and grasps it; the direction of this final pounce is not correlated with wind direction. I chase it off again, and displace the bee approximately 10 cm. The wasp once more flies round in wide loops, then approaches in the described manner against the wind. It is chased off a third time, with essentially the same result.

Experiment 2 In order to identify which sensory modality guides the wasp during its upwind retrieving flight to the bee which it has dropped, I apply the simple test used by Knoll (5). After having robbed a returning wasp of its bee as described, I put the bee in the bottom of a test tube, which was then put out near the nest site, oriented at right angles to the direction of the wind. In some tests the concentration of bee-scent was increased by providing more than one bee in the test tube; in those cases I took care to put one of them approximately 1 cm away from the others, so that it could be visually recognised as one single bee. If the wasp were guided by visual stimuli it should of course make straight for the bee and bump into the glass; if by chemical stimuli it should go into the mouth of the tube.

This was done with 11 wasps in all. Seven of these tried to reach their prey on 11 occasions in all; on each occasion they entered the tube through its entrance and no visually controlled response was observed.

The remaining 4 wasps did approach from downwind but changed their behaviour suddenly when still approximately 25 cm away, when they started to zigzag wildly, their body axis pointing to the test tube, but failed to approach any nearer. I soon realized that on all these occasions the wind happened to come from the direction of the sun, and that, from the point of view of the approaching wasp, the glass vial glittered brightly. My supposition that this prevented further approach was confirmed when I repeated the test 6 hours later with two of these wasps; both found the test tube entrance without any hesitation, and took the bee lying nearest to the entrance.

These tests showed not only that the retrieving of a dropped bee is guided by olfactory stimuli, but also that the wasp responds positively to a wide range of concentrations of bee-scent.

Before proceeding further I checked whether the retrieving response was confined to the wasp's own particular bee, or whether it would retrieve any dead honey bee.

Experiment 3   After having made a returning wasp drop its bee, I offered it its own bee together with a bee taken from another wasp, and a bee killed with cyanide, all arranged in a line at right angles to the wind, and at intervals of a few cm. In all, 7 wasps took their own bee 10 times, the bee taken from another wasp 13 times, and the cyanide-killed bee 9 times. I concluded that the wasp did not show a preference for its own bee; this was also indicated by the fact that during such experiments neighbouring wasps would sometimes take a bee before my own wasp had retrieved it. This was particularly frequent on windy days, and such 'robber wasps' often showed the 'pouncing' behaviour described above.

Experiment 4   To isolate the scent stimulus in another way, I cut dry, withered pine or heather twigs into roughly bee-sized pieces. Some of these twigs were then shaken vigorously in a glass tube with freshly killed bees, and one of these scented twigs was laid out together with one or two similar but untreated twigs. I presented these twigs to 10 wasps, who responded 24 times in all; in every case the wasp approached the scented twig as it would approach a bee and ignored the other twigs. The following entry provides some details of a test in which some other factors were varied at the same time:

18.7.34. 16.30 hrs. A *Philanthus* carrying a prey lands by a sand patch situated on a dark, lichen-covered background. I chase it off and, in place of the bee it has left behind, I put out a bright, white, bee-scented twig and two dark, unscented twigs, thus presenting the scent attached to a dummy of the wrong colour, which in addition contrasted less with the background than did the two non-scented dark twigs. The wasp returns, approaches the scented twig from downwind, flying at first, then walking, and grasps it, taps it vigorously with its antennae, drops it again, seizes it once more, but ultimately abandons it. I chase the wasp off, take the bright twig away, and put out a dark, scented twig some cm downwind from one of the unscented dark twigs, so that, if it were guided by scent, it would have to pass by the unscented dummy. Without any noticeable hesitation the wasp passes the latter and grasps the scented twig with all six legs, clamps it to its body as if to fly off with it, then drops it. I

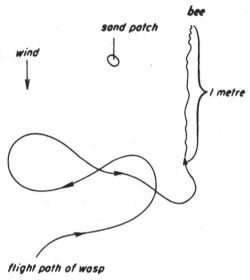

Fig. 25. Explanation in text.

repeat this 3 more times; each time the non-scented twigs are ignored and the scented one taken.

While it is therefore clear that a bee dropped near the nest entrance can be found by scent alone, the fact that the scented twigs were not treated in exactly the same way as a bee shows that the wasp can distinguish between the two. It does so in part by means of visual cues; for instance the dark scented twigs were often stung, but this never happened to the white scented twig. My experience with the test tubes glittering in the sun also points to the power of visual stimuli, and more information will be given later on. This raised the question of whether a bee can be discovered by means of visual cues alone. Information about this is given in:

Experiment 5 A bee killed by a *Philanthus* was placed for two days in 96 per cent alcohol, which was renewed several times, and then dried in the sun. While naturally looking different from an untreated bee, it nevertheless resembled one much more than did a dark twig. It was offered in the usual way to 10 wasps, all of whom failed to respond. When however I shook the same 'de-odourised' bee with fresh *Philanthus*-killed bees, it was immediately taken by the first wasp it was presented to.

For the retrieving of a dead bee, smell is therefore the essential cue, and vision plays at best a subsidiary part. As already mentioned, the scent of a bee was picked up from a distance of approximately 1 m; in rare cases from up to 2 m (Fig. 25).

As I have already stated, it often happened on windy days that wasps roaming over the colony responded to bees lying on the ground when these stirred in the wind. I also noticed on a number of occasions that such 'loiterers' would follow a wasp descending with a freshly captured bee; they hovered behind it at some 15 cm distance, and often swooped down on it, when both wasps would fall to the ground. Only rarely however did the owner release its bee, and the would-be robber flew off at once. Whereas such a robber wasp would initially follow its intended victim closely, as if, like a Tachinid fly, it were joined to it by an invisible thread, the final approach was in the form of a very fast 'pounce'. Similar 'pouncing' attacks were seen when one of my experimental bees was moved by the wind, or when it was lying on one of the light sand patches. This led me to believe that the final pounce might well be guided by visual stimuli. As we shall see, this final pounce, though observed rarely during retrieving, is a regular part of the wasp's behaviour when catching a live bee, and I therefore paid special attention to its occasional occurrence in retrieving. This led me to arrange:

Experiment 6 After a few pilot tests I set up the following arrangement. A horizontal piece of thin black yarn was tied loosely between two sticks stuck upright in the soil, to which a bee, made odourless in 96 per cent alcohol, was attached in the middle, approximately 30 cm above the ground. About 3 cm upwind from this bee a fresh, *Philanthus*-killed bee was strung up in the same way (Fig. 26). As the threads were not tightly stretched, both bees could be gently moved by the wind. Thus any wasp approaching the scent-bearing bee upwind would have to pass the odourless bee. If the wasp's final pounce were visually oriented, the wasp could be expected to pounce on this rather than on the normal bee.

The responses were not uniform. Sometimes a wasp approached

THE ANIMAL IN ITS WORLD

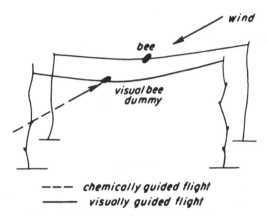

--- *chemically guided flight*
——— *visually guided flight*

Fig. 26. Experiment 6: arrangement.

quite normally until it was approximately 15 cm from the de-scented bee. It then stopped and hovered in mid-air, copying exactly the movements of the strung-up bee, but not approaching any nearer, and finally flying off. This was seen especially in strong wind, when the bee moved vigorously. A second type of response was indistinguishable from the normal retrieving behaviour: the wasp approached slowly upwind, and grasped the *Philanthus* bee, bypassing the odourless one. On such occasions however no pounce was observed. This happened on very quiet days, when neither dummy moved, and also when the wind was light, and in addition had changed direction between the setting-up of the test and the response of the wasp (Fig. 27). Approaching upwind, such a wasp would not find the odourless bee exactly in its flight path. Since in neither of these situations pouncing occurred, these observations had no bearing on my problem. However, a real pounce was observed on 12 occasions. These occurred when the bees moved slightly, and when the wasp found the odourless bee exactly in its flight path. Without exception, these pounces were aimed at the odourless bee, thus proving that the pounce was indeed oriented by visual cues. The following entry is typical:

3.8.34. 12.25 hrs. A de-scented bee is presented 4 cm upwind from a fresh *Philanthus*-killed bee. The wind is light and both dummies move slightly. The wasp approaches from the lee side, zigzagging and proceeding against the 'scent-stream', stops for an instant when about 10 cm from the odourless bee, and then pounces on it. It grasps it for a split second, then flies off. It has flown hardly 60 cm when it happens once more to be caught in the scent-stream.

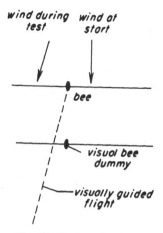

Fig. 27. Explanation in text.

The wind, which had been variable, has now turned through some 30 degrees, and the wasp's approach is clearly aimed at the scented bee. When it has come to within some 30 cm the wind momentarily drops, and the wasp slowly approaches and takes the scented bee without pouncing.

The overall interpretation of the data obtained so far was that a bee which has been accidentally dropped is retrieved mainly by scent and that visual stimuli are ineffective in eliciting the pounce except when the bee either moves gently or contrasts strongly with the background.

I had initially hoped to be able to analyse the full hunting behaviour by watching the wasps in their nesting area, concentrating on those occasions when they retrieved an already motionless bee. I now began to realise that the slow, chemically oriented approach to such a bee might not reflect the normal hunting behaviour. For one thing such a slow approach to a normally extremely mobile prey would not seem to be very effective; for another, the pounce seemed not only to be efficient, but had appeared under circumstances that might be typical of wasps in the hunting field. This was borne out when I moved to the wasps' hunting grounds, but as we shall see it turned out that in some respects the complete hunting behaviour differed fundamentally from retrieving.

The Hunting Sequence Proper

In the second half of August I decided to follow the wasps to their hunting area, a stretch of heath where unlimited numbers of bees and numerous other insects were foraging on the flowers of *Calluna* and *Erica*.

After a few days' observation I found that my fears, that on this wide stretch very few wasps would ever stray near enough for observation, were unfounded, and that the actual hunting could be observed often enough for a pattern to become evident. The bees were captured when gathering nectar on the flowers. The wasps, appearing each sunny morning approximately 30 minutes after the bees had begun to forage, roamed over and round the *Calluna* plants, flying at most some 60 cm above the ground, and occasionally settling down on bare patches of ground. Not all the bees foraging near such a wasp were attacked; I saw wasps fly past 4 or 5 plants in succession without responding to the several bees that were foraging on each of them. They seemed to ignore the latter as long as they were on a flower quietly collecting nectar, but a bee flying slowly from spike to spike was noticed from distances up to approximately 30 cm. Such bees were approached with remarkable speed; the wasp would fly to a position some 10–20 cm away from the bee, hover on the spot and, after 1 or 2 seconds, pounce down on to it so quickly that I had difficulty in following it. The next moment the wasp, having grasped the bee between its legs, tumbled down through the jungle of *Calluna* twigs, and still holding on to the bee, fell to the ground. Here it stung the bee in the manner already described, and then proceeded to press it with force against its own body, licking up the nectar as it appeared from the bee's mouth. Bees were detected by wasps which were flying as well as those which had settled. In the latter case the wasp always turned its whole body towards the bee before flying up; since this happened irrespective of the direction of the wind it suggested visual orientation. However in the 50 or so complete observations made in this initial period, the hovering-on-the-spot, which always preceded the final pounce, was done in a position exactly downwind of the bee.

Still hoping to save time by not having to wait until a hunting wasp happened to come within sight, I once more tried to elicit the full hunting behaviour, as I had now seen it, in captivity. I now put up, in the hunting area, a flight cage measuring approximately 50 × 50 × 50 cm, which I put over some *Calluna* plants, and into which I released one or more wasps and several bees. But even in this setup the wasps tried to escape rather than to hunt, and, as observed previously under the glass dome (13), would capture and sting a bee only when one accidentally touched their antennae. I therefore had to conduct my experiments entirely in the open. For this I selected a relatively narrow bottleneck in the vast heath, through which wasps flew regularly back and forth between the nesting and the hunting areas. To my delight it proved possible to carry out several series of experiments on wasps which were actually hunting.

Experiment 7 A bee which I had killed with cyanide was hung from a horizontal thread strung up loosely between two vertical twigs, as described in the previous section. A light wind usually moved it slightly. It was not long before it was noticed by a passing wasp, which quickly took up a position 10 cm downwind of the suspended bee, hovered for a while and then pounced. It actually grasped the bee, stung it and then, after having struggled for a while in an attempt to carry it away, pressed the bee tightly against its body, licked up the nectar and tried to make off with it—of course without success. Having thus at last succeeded in eliciting the full hunting sequence I could proceed with the analysis of the parts played by visual and other stimuli in the normal hunting behaviour.

For this I used the following 'bee dummies':

Experiment 8 A bee freshly killed with cyanide was always used as a control; it was always provided in close proximity to the other dummies used;

Experiment 9 A bee made odourless in alcohol in the way described above;

Experiment 10 A twig, approximately bee size, cut from dry, withered *Calluna* stems, similar to those already used in the retrieving experiments;

Experiment 11 A similar twig which had been given bee-scent by shaking it for some time in a glass tube containing dead bees.

All these dummies were offered on quiet but not completely windless days, and all were strung up as in experiment 7, but I now offered several different dummies spaced along one thread.

Experiment 12 A 3 cm-wide glass vial, which was open at one end and contained a few bees freshly killed with cyanide, was put among the *Calluna* plants, but was screened from vision by a small fence of pine twigs downwind of the tube;

Experiment 13 A bee killed with cyanide was tied to a dead *Calluna* twig;

137

Experiment 14 A similar bee was tied to a flowering *Calluna* spike;

Experiment 15 A de-scented bee which was given bee scent by shaking it in a tube with freshly killed bees was either tied to a flowering *Calluna* twig or suspended from a thread.

In most cases it proved impossible to set up and observe all these experiments at the same time and at any one session I used 4 or 5 categories of dummies, in irregular combinations. Even so, it was easy to miss the arrival of a wasp, and my thanks are due to Miss T. Kooistra and to my brother L. Tinbergen for assisting me in the observations. The number of dummies in any one category varied, but the results were made comparable by noting, for each category, the total number of 'bee-hours' and by expressing the 'response rate' to each type of dummy by the relationship: 'number of responses observed' divided by 'number of bee-hours'. In Table 11 the figures in parentheses indicate the number of responses per 100 bee-hours.

Table 11. *General survey of experiments 8–15*

Experiment No.	'Bee Hours'	Hovering	Pouncing	Grasping	Stinging	Capture
8 (Dead bees)	153	40 (*26*)	5 (*3*)	17 (*11*)	11 (*7*)	29 (*19*)
9 (Alcohol-treated bees)	88	18 (*20*)	8 (*9*)	—	—	—
10 (Scentless twigs)	101	14 (*14*)	1 (*1*)	2 (*2*)	—	—
11 (Scented twigs)	119	40 (*34*)	3 (*3*)	20 (*17*)	11 (*9*)	—
12 (Glass vial with bees)	22	—	—	—	—	—
13 (Bees tied to twigs)	70	—	—	—	—	—
14 (Bees tied to flowering *Calluna*)	36	—	—	—	—	—
15 (Re-scented alcohol-treated bees)	13	3	—	6	2	—

The hunting behaviour of the wasps was not always shown in full: a dummy might be noticed by a wasp but not approached, grasped, stung, or the sequence might break off at an even later stage. It was clear that the hunting behaviour consisted of a chain of separate responses, each of which was dependent on special stimuli. The successive steps of the whole sequence were named as follows:

By 'hovering' I mean the Syrphid-like flying-on-the-spot during which the wasp faces the bee or dummy from between 5 and 20 cm downwind. 'Pouncing' was the term given to the extremely rapid flight leading to touching the bee, without actually grasping it. Although pouncing happened too fast for the eye to follow, one could see whether or not the wasp had touched the bee by the latter's slight movement. 'Grasping' is largely self-explanatory; it is used for those occasions when the wasp took hold of the 'prey' without however making an attempt at stinging. 'Stinging' includes at least an attempt to sting (clearly recognisable by the curving of the abdomen), but does not include pressing, sucking and licking up the nectar. By 'Capture' I mean the entire sequence, which includes all previously mentioned responses as well as an attempt to carry off the 'prey'— which of course was always prevented by the dummy's being firmly fixed. The table always mentions the act last seen. This means that if, for instance, I recorded for any individual wasp '2' for stinging and '6' for grasping, I had seen it pounce 8 times. The table therefore lists the acts after which a sequence broke off.

It can be seen at a glance that bee-scent alone never attracted a wasp (experiments 12, 13 and 14). Here, as in the section on re-trieving, one could argue that the concentration of bee-scent offered in experiment 12 was too high, but even so the interesting fact remains that a retrieving wasp did respond to the concentrated bee-scent whereas a hunting wasp did not. Since the hunting experiments were done later in the season than the retrieving experiments, I repeated the retrieving test a few times in late August in the nesting area, and obtained the same positive responses as before. On the other hand, variations in the concentration did not affect the pouncing, for the scented twigs of experiment 11 gave off a distinctly stronger scent than a normal bee and yet were often pounced upon and even stung.

While the initial approach of a wasp might often go unnoticed, the hovering could not be missed, nor misinterpreted since the wasp maintained such a precise orientation with respect to the dummy. The relatively high values for hovering even in experiments 9 and 10 show clearly that the bee is discovered visually. The fact that the unscented dummies have a lower score for hovering than the scented dummies is undoubtedly due to the fact that a wasp hovers only very briefly in front of an unscented dummy; we are convinced that a number of very brief responses (those under $\frac{1}{2}$ second) have been overlooked under the difficult field conditions; and we did not record doubtful responses. The hovering in front of a scented dummy might last from appr. 1 second to as long as 8 seconds.

The failure of wasps to respond to the bees tied to twigs, whether dead or flowering (experiments 13 and 14) further indicated that, in order to be seen in the first place, the bee has to be moving; this was also clear from the fact that almost all hovering responses were seen when there was a light wind. Once I had established this, I used to make all dummies attached to one thread move slightly by gently pulling two threads attached to the vertical twigs in as standardised a way as possible, whenever I saw a wasp nearby; by thus raising the overall response level I could obtain larger figures for the responses which followed hovering. It was in this way that I was able to establish that the way the wasp responded to dead bees (experiment 8) and scented twigs (experiment 11) on the one hand, and to odourless bees (experiment 9) and unscented twigs (experiment 10) on the other was strikingly different: the odourless dummies were rarely pounced upon or grasped, and never stung.

The few exceptions are explained below. Obviously hovering downwind of the bee enables the wasp to test the scent of any insect that has been discovered visually.

I have already mentioned that when a retrieving wasp pounced onto the bee, as happened occasionally, it was guided by vision. Incidental observations on the hunting wasps showed that this was true here as well. A wasp hovering in front of a real, free-flying bee (of which there were of course many to be found foraging right among my dummies) would often lose it when it crawled from flower to flower and disappeared behind a spike. On such occasions the wasp might turn towards the near-by unscented dummies and pounce upon one of them, particularly if they were being moved by the wind. The majority of 'pounces' on unscented dummies occurred under such circumstances.

We must therefore conclude that, while the pounce was *elicited* by the correct scent, it was *oriented* in space by means of visual cues. However, on impact an unscented dummy was immediately released, showing that for the next stages (stinging, etc.) the chemical properties were again of importance.

It might be objected that the bees used in experiment 9 (which under the alcohol treatment had changed their shape to some extent), were ignored because of their visual rather than their chemical properties. That this was not the case was shown by experiment 15, which for obvious reasons was not considered worth repeating frequently.

The wasps often tried to sting a scented twig. However, they could not penetrate into the hard wood, however much they tried, and it was after such stinging attempts that the twigs were abandoned.

Taken together, the results give a fairly clear picture of the sensory control of the successive stages of the hunting sequence (Fig. 28). The bee is detected visually, and for this it merely has to have roughly the size and shape of a bee, and must preferably be moving slightly, as a bee which is actually foraging does once it has alighted. The wasp detects such bees either when it is itself in flight or sitting on the ground. It responds by taking up a hovering position 5–20 cm downwind, when

140

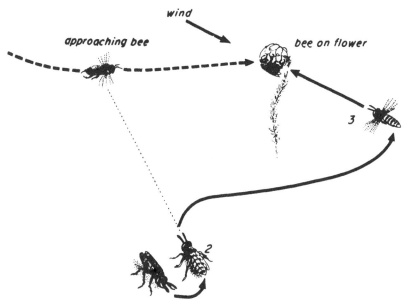

Fig. 28. The hunting sequence.

it tests the bee's scent. Bee-scent elicits pouncing, but the orientation of the pounce is visual—as its extreme speed had of course suggested all along. For grasping and stinging to occur the olfactory stimulus must still be present. Unknown but undoubtedly tactile or kinaesthetic stimuli enable the wasp to distinguish between a bee and the hard twigs, and these stimuli obviously determine whether or not stinging or a stinging attempt shall be followed by 'pressing', feeding, and taking the prey home.

A few entries give an impression of the kind of observations on which the table is based; they also provide some additional, qualitative evidence:

24.8.34.   9.35 hrs. A *Philanthus* discovers a foraging bee and hovers in front of it; the bee however flies rather quickly from flower to flower, pausing only briefly on each *Calluna* spike. The wasp follows the bee about, staying downwind as accurately as possible. When the bee disappears for a moment into the heather behind one of the odourless twigs, the wasp pounces on this twig, grasps it, and even holds it for about $\frac{1}{2}$ second, but it does not sting. The bee moves into sight again; the wasp hovers once more near it, but now the odourless twig is exactly half way between it and the bee, and the wasp pounces once more on the twig, treating it as before. The bee now

141

happens to fly rather fast to the next *Calluna* bush; *Philanthus* loses sight of it and flies away.

17.8.34 15.27 hrs. A *Philanthus* settles on a *Calluna* twig. A medium-sized *Bombus terrestris* worker settles on the heather 32 cm away. The wasp quite clearly fixates it, turning its entire body towards it and following all its movements by movements of its head. It then flies up, hovers approximately 15 cm downwind; after 1 second it flies off. The bumblebee was considerably larger than a honeybee.

25.8.34. 11.59 hrs. A *Philanthus* sitting on the sand responds to a small bee (*Halictus* sp. ?) foraging on *Calluna* some 15 cm above it. It flies up, hovers for $\frac{1}{2}$ second downwind; just at this moment the bee is practically still, and the wasp pounces on a small, withered twig approximately 3 cm from the bee, then flies off. The bee was approximately half as long, and much more slenderly built than a honeybee.

26.8.34. 9.34 hrs. Four scented twigs (D_1, D_2, D_3, and D_4) are hanging, with appr. 5 cm intervals, on a horizontal thread at right angles to the wind. A wasp hovers several seconds downwind of D_4, then switches to D_3, then to D_2, each time standing between 5 and 10 cm from the dummy. The switch from one dummy to the next is characteristic (Fig. 29): each time the wasp turns suddenly and exactly towards the next dummy and then moves smoothly to the new position facing the new dummy, going through the positions 1 to 4, and all the time exactly facing the new dummy. Here she hovers on the spot until her attention is obviously caught by the neighbouring dummy, when the sequence repeats itself.

Observations such as these show that under natural conditions responses to insects visually similar to bees are not rare. All in all, I observed 47 full hunting sequences in response to free-flying honeybees; 'numerous' hovering responses and even 11 pounces on various flies, bumblebees, etc. Since even these were never stung I conclude that not only the pouncing but also the final decision to sting depends on a scent specific to the honeybee.

It also became clear that, although the main function of the hovering is the chemical testing of the insect seen, the wasp's eyes are not totally 'switched off'; on the contrary, the wasp keeps following the bee with remarkable accuracy, and also switches easily to another insect when the bee happens to disappear from sight, even though it remains within 'scent distance'. Also, the wasp does not pounce as long as the bee is moving vigorously. All the same, the switch in the use of sensory input from one stage of the hunting sequence to the next is a striking phenomenon. For instance, although even a hunting wasp can smell a bee from at least 15 cm, no single wasp ever responded to the sources of bee-scent presented in experiments 12, 13 and 14, despite the fact that I saw a number of wasps pass only a few cm downwind of these dummies. Even more striking is of course the fact that a wasp retrieving a lost bee on the nesting ground can smell it from up to 2 m away.

I also considered the possibility that a hunting *Philanthus* makes use of auditory stimuli. I have never offered an acoustical stimulus in isolation, but my observations rather suggest that sound is not involved. Bees disappearing from sight often buzzed audibly, but I never saw a wasp respond to this, whereas responses to bees which were visible but were simply crawling around quietly were frequent.

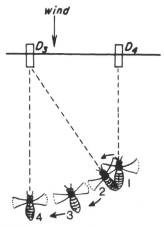

Fig. 29. Visual switch from scented dummy D4 to scented dummy D3.

A digging wasp can not be disturbed by loud noises, whereas even slight sudden movements of the observer scare it off. I have therefore no indication that *Philanthus* can hear at all.

Tactile stimuli probably play a part in determining whether to sting a prey or not. Although the antennae undoubtedly possess sensitive touch receptors, touch alone does not elicit any part of hunting; even under the glass dome, when the wasps were not inclined to hunt, the antannae have to be touched by an object carrying bee scent.

Summary

Philanthus is able to retrieve its prey after it has dropped it when disturbed upon its return to the nest. In this situation it does not remember the exact location of the bee, but roams over an area of 1–2 sq. m. It finds the bee with the aid of olfactory stimuli; under favourable circumstances, it can smell such a bee from up to 2 m. A wasp does not distinguish between its own bee and those lost by others. When the bee is either moved by the wind or contrasts vividly with the background, the wasp makes a quick dash or 'pounce' from appr. 10 cm; this pounce is guided by visual cues; and so the wasp occasionally pounces on twigs or pebbles of roughly the same size as bees.

By contrast, a wasp hunting for live bees is not attracted by the scent of bees, but it responds visually to a wide range of insects differing in size and details of shape. Honeybees are seen from approximately 30 cm. Responses to various flies and bumblebees are not rare, but they break off when the wasp positions itself, as it invariably does, some 5–15 cm downwind of the insect—hovers on the spot and occasionally follows the insect about. Unless the insect gives off the scent typical of a honeybee, the sequence is broken off after hovering. Honeybee-scent *elicits* the 'pounce', but this very rapid movement is *oriented* visually. When it touches the prey, the wasp once more tests it for bee scent. The part played by vision and olfaction therefore varies not only with the motivation (retrieving of a dead bee, and capturing of a live one respectively), but also, within the hunting sequence, from link to link of the action chain. It seems possible that Molitor's results, showing that a wasp digging at its burrow will accept a variety of insects, indicate that a wasp 'in digging mood' responds to yet another set of stimuli. Chemically determined food preferences have been reported for other insects as well (Verschaffelt, **14**; Von Frisch, **3**; Murr-Danielczick, **11**; Von Stein-Beling, **12**). The hunting of *Philanthus* (initiation by visual stimuli followed by olfactory checking) is similar to the feeding behaviour of the Hymenoptera specialising on flowers, such as the honeybee (Von Frisch, **1**, **2**; Kroll, **5**) and bumblebees (Kugler, **6–9**).

Responses to either acoustical or tactile stimuli alone were never observed; there is no evidence that *Philanthus* can hear at all.

REFERENCES

1 FRISCH, K. VON (1914). 'Der Farbensinn und Formensinn der Biene', *Zool. Jb.*, **35**.

2 —— (1919). 'Über den Geruchsinn der Biene und seine blütenbiologische Bedeutung', *Zool. Jb.*, **37**.

3 —— (1926). 'Vergleichende Physiologie des Geruchs- und Geschmackssinnes', *Handbuch der normalen und pathologischen Physiologie*, Bd. **11**.

4 GRANDI, G. (1930). *10° Contributo alla conoscenza biol. e. morfol. degli imenott. mellif. e predatori*, Bologna (after Molitor).

5 KNOLL, F. (1926). 'Insekten und Blumen', *Abh. Zool. Bot. Ges. Wien*, **12.**.

6 KUGLER, H. (1930). 'Blütenökologische Untersuchungen mit Hummeln', *Planta (Berl.)*, **10.**

7 —— (1932). 'Blütenökologische Untersuchungen mit Hummeln III', *Planta (Berl.)*, **16.**

8 —— (1932). 'Blütenökologische Untersuchungen mit Hummeln IV', *Planta (Berl.)*, **16.**

9 —— (1933). 'Blütenökologische Untersuchungen mit Hummeln VI', *Planta (Berl.)*, **19.**

10 MOLITOR, A. (1934). 'Neue Beobachtungen und Versuche mit Grabwespen V', *Biol. Zbl.*, **54.**

11 MURR-DANIELCZICK, L. (1930). 'Über den Geruchssinn der Mehlmotten-schlupfwespe *Habrobracon juglandis* Ashmead', *Z. vergl. Physiol.*, **11.**

12 STEIN-BELING, I. VON (1934). 'Über den Ausflug der Schlupfwespe *Nemeritis canescens* Grav. und über die Bedeutung des Geruchssinnes bei der Rückkehr zum Wirt', *Biol. Zbl.*, **54.**

13 TINBERGEN, N. (1932). 'Über die Orientierung des Bienenwolfes (*Philanthus triangulum* Fabr.)', *Z. vergl. Physiol.*, **16**

14 VERSCHAFFELT, ED. (1910). 'De oorzaak der voedselkeus bij eenige plantene-tende insecten', *Verslag. Kon. Acad. Wetensch.*

(From the Zoological Laboratory of Leiden University)

4

On the Orientation of the Digger Wasp *Philanthus triangulum* Fabr. (1938):

III. Selective Learning of Landmarks[1]

1. Introduction: the Problem

The females of the digger wasp *Philanthus triangulum* Fabr. dig out deep burrows in sandy or loamy soil (Fig. 30), ending in a number of cells or chambers in each of which they lay an egg, which is then supplied with a number of immobilised honeybees. These bees are caught on their feeding grounds, which are often far from the *Philanthus* nests. Since each burrow is provisioned with 10 or even more bees, each wasp has to home to its nest a number of times. In the course of the summer, several nests are made and provisioned in succession. In an earlier paper (15), the senior author demonstrated that the wasps find their burrows with the aid of landmarks surrounding it. These landmarks are perceived visually, and are known to the wasps by experience. This was shown by displacing landmarks in the absence of the wasps, which upon return would alight on the place which bore the correct relation to the landmarks. However, a number of observations remained unexplained: on some occasions the displacement of landmarks which to us seemed highly conspicuous did not mislead the wasps. Similar, seemingly contradictory observations have been made by various authors on other Hymenoptera (Fabre, 2; Bouvier, 1; Verlaine, 16), and it was observations of this nature which have in the past led to suggestions that the homing behaviour might not depend entirely on the use of landmarks, but might involve the use of stimuli emanating directly from the nest itself.

We found however that if we made drastic changes to the im-

[1] We wish to express our thanks to Professor Dr C. J. van der Klaauw for helping us in many ways, to Ir A. E. Jurriaanse for his kindness in allowing us to work on his land, and to Messrs D. J. Kuenen and A. F. H. Besemer for their help in some of the experiments.

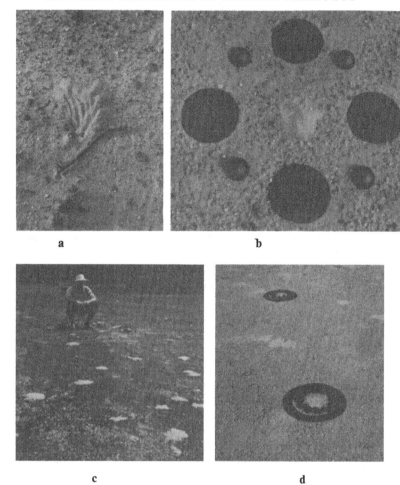

Fig. 30. Top left: *Philanthus* burrow on sandy soil. Top right: training situation with hemispheres and large flat discs. Bottom left: *Philanthus* burrows on lichen-covered soil. Bottom right: training situations with blackened pine cones and flat rings.

mediate surroundings of a *Philanthus* burrow, for instance by irregularly scattering *all* objects found within a radius of, say, 1 m from the nest, the homing wasp would be completely disoriented. It would search for a long time but would not find the entrance to the burrow, although this had not itself been disturbed. Such disorientation occurred not only in wasps which we had previously trained with

147

conspicuous landmarks, but equally in the 'owners' of natural, undisturbed burrows, of which the surroundings often seemed to have few conspicuous features. Bouvier (1) describes a similar observation on *Bembyx labiatus* Fabr. This apparent contradiction called for a further investigation.

A chance observation led us to the solution. In the paper mentioned earlier, Tinbergen had shown that the digger wasps could not be made to accept large flat sheets of paper as landmarks, even when these were placed quite close to the nest. Yet groups of pine cones, covering areas of the same size, were used after quite short periods of training. The pieces of paper in question, 9 × 12 cm in size and bright blue, red or yellow in colour, were to human eyes more conspicuous than the drab pine cones. The most likely interpretation was that the wasps' preferential choice of landmarks must differ radically from our own. Hence a study of the characters by which a wasp chooses its landmarks looked promising, the more so as Hertz (5) had previously discovered and analysed such preferences in honeybees.

We studied this problem in some detail during the summers of 1934, 1935, 1936 and 1937. As before, we worked in a stretch of open sand, with patches of lichens, mosses and isolated tufts of grass, and an occasional pine tree.

2. Method

When we first tried to compare the effectiveness of various types of landmarks we placed objects of one type near one group of nests, and objects of another type near another group; returning after a standard period of time, we shifted the objects at each nest in order to record possible differences in the effectiveness of the two types of landmarks. Unfortunately the results of these experiments were inconclusive because, as it turned out, the level of training achieved did not depend simply on time, but rather on the number of flights into and out of the nest, and on other factors which varied between individuals. After some preliminary experiments, we chose the following method:

The two types of objects to be tested, (a) and (b), were placed in roughly similar positions close to a *Philanthus* burrow so that the wasp, in its regular flights in and out of the burrow, was exposed equally to both. As soon as the wasp could be misdirected by our shifting of the whole configuration (as described by Tinbergen, 15) and so showed that training had been successful, we prepared two dummy nests. These were usually placed in such a manner that one was to the right and the other to the left of the original nest as seen by a wasp alighting against the wind, as is their custom. We then placed all (a) objects near one, and all (b) objects near the other of the two dummy nests. As soon as a wasp had alighted at one of these nests we chased it away before it had time to start digging, in this way forcing it to repeat its choice. We were careful to see that it flew far enough away to be forced to make a completely new choice. This need not be far in view of the poor resolving power of its eyes (see Tinbergen, 15). Quite often we could judge from the behaviour of the wasp that we had been successful in this: frequently the insect would restart its descent from many metres distance, fly to and fro between the two dummy nests and would not alight again before it had hovered over both several times. In this way we were usually able to force a homing wasp to make between 5 and 20 decisions. In exceptional cases we succeeded in achieving an even longer series. It was impossible however to force each wasp to make a constant number of choices—some wasps managed to slip in prematurely, while others would give up and disappear altogether for hours.

The way in which we arranged the training varied. Kruyt always used two concentric rings, which meant that, during the training, one of the two types was always further away from the entrance to the nest than the other. Since, as will be shown later, distance from the nest is a factor affecting results, a preference for (a) or (b) could only be determined by training to two situations: one in which (a) was outside and (b) inside, and one in which their position was reversed. This meant that during the training each type of landmark had to be used in two sizes. When it came to the actual tests the effect of size and distance was eliminated by offering either both (a) and (b) 'small', or (a) and (b) 'large' (Fig. 31).

The drawback of this time-consuming method of testing was offset by the fact that training was accomplished more quickly than with the arrangement favoured by Tinbergen. In view of *Philanthus'* extréme dependence on warm weather for its activities, and our need for prolonged series of tests, this was an important consideration.

In order to overcome just this disadvantage of having to present our landmarks at different distances, Tinbergen used a number of smallish objects which he arranged in a single circle around the entrance of the burrow so that they all were at exactly the same distance from it. Each circle consisted of 4 to 6 objects of type (a) and an equal number of objects of type (b) placed alternately (Fig. 32). During testing, one set of objects was again placed round a sham nest to the right, the other round one to the left of the original nest. The fact that these test circles contained only half the number of individual objects present in the original circle to which the wasp had

149

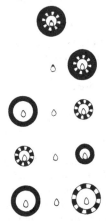

Fig. 31. Kruyt's method of concentric rings. Top: training situation; in descending order: first check that training had been successful; three successive tests.

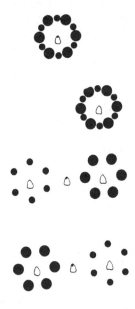

Fig. 32. Tinbergen's method of alternating landmarks. Top: training situation; in descending order: three successive tests.

been trained turned out to be of little importance, as will be shown later. Although training to the less conspicuous small objects tended to take longer than training to the more substantial rings employed by Kruyt, these small objects proved in certain respects more suitable.

With both arrangements we naturally sought to eliminate position effects by switching the position of (a) and (b) from time to time.

Experiment 1   To see whether the method would succeed we trained some wasps to a flat ring, uniformly black in colour, and to a circle of pine cones, also dyed black (Fig. 30: bottom right). The wasps chose:

Wasp No.	Flat ring	Cones
1	0	6
2	0	11
3	0	11
4	0	14
5	0	8
6	0	4
7	0	8
8	0	6
9	0	8
	0	76

This showed that our method was workable. It should however be mentioned here that such a definite preference for one of two alternatives was not found in all our experiments. And even when it happened, it was often possible to show that the landmarks which were ignored when offered together with the others were not completely ineffective: when the 'strong' type was removed, the wasp would choose the sham nest with the previously ignored 'weak' landmarks, and not the real nest.

In most cases the preference for one type of object over another was less clear cut; with few exceptions, the less attractive landmarks were chosen several times, and even in some cases only slightly less often than the more effective ones. That is why long series were usually required. Another reason why it was important to have long series of experiments was a peculiarity of which more will be said later: where landmarks did not differ from each other very greatly, any preference would show only after prolonged training, lasting several days. All our experimental animals had of course memorized the surroundings of their nests before we started their training, and the landmarks we added had therefore in every case to erase the earlier, natural learning. Hence, the time required for training varied with the effectiveness of the landmarks offered. Once a wasp was trained we could expect to carry out 1 to 3 tests per day each with 5–20 observations per nest, provided the weather held. We therefore quite frequently required a number of days to get a reasonably satisfactory series, and a great many experiments had to be terminated because the weather broke.

It was the observation that it was easier to train the wasps to use pine cones than flat discs which first drew our attention to the differences between these two types of objects. Initially we set out to test the effectiveness of the following parameters: the solidity of pine cones; the three-dimensional structuring of their surfaces; and the two-dimensional structuring caused by the projection of this structuring, i.e. the shadows cast upon the ground and on themselves. While, as it turned out, these were all used by the wasps, we found that even more parameters were involved.

3. Two-dimensional Patterning

We soon discovered that training to flat landmarks was not totally impossible—it merely took longer than training to, say, pine cones. The effect of two-dimensional patterning could therefore be tested with flat landmarks. As at this stage, Kruyt's method was in practice simpler, we trained our wasps exclusively with flat rings. We worked with rings that were either uniformly black, or uniformly sand-coloured ('yellow'), or patterned black and yellow.

Experiment 2   A plain black ring versus a ring with 4 black and 4 yellow sectors of the same size (Fig. 33).

Wasp No.	Patterned	Black
10	8	5
11	4	0
	12	5

Experiment 3   A plain black ring versus a ring with 8 black and 8 yellow sectors (Fig. 34).

Fig. 33. Experiment 2.

152

Fig. 34. Experiment 3.

Fig. 35. Experiment 4.

Fig. 36. Experiment 5.

Wasp No.	Patterned	Black
12	11	3
13	3	4
14	8	1
15	4	0
	26	8

Experiment 4 A plain black ring versus a ring with 16 black and 16 yellow sectors (Fig. 35).

Wasp No.	Patterned	Black
16	11	3
17	18	0
	29	3

After these tests, which showed that rings with black and yellow sectors were preferred to plain black ones, we offered rings with black and yellow check patterns, i.e. the two colours alternating not only in sectors, but also concentrically.

Experiment 5 One ring was plain black, the other had black and yellow sectors in two concentric rings (16 of each colour) (Fig. 36).

Wasp No.	Patterned	Black
18	5	8
19	9	4
	14	12

Fig. 37. Experiment 6.

Experiment 6 One ring was plain black, the other had even smaller checks than in experiment 5, namely 64 sectors each subdivided 4 times radially (Fig. 37)[1]

Wasp No.	Patterned	Black
20	6	0
21	3	0
22	2	2
23	8	0
24	0	6
25	18	3
26	14	8
27	6	5
28	1	5
	58	29

All these experiments were carried out on bare sand. Because the yellow colour we used was to our own eyes virtually the colour of sand we initially used a black ring as the unpatterned alternative rather than a yellow one, so that we could exclude the possibility that the patterned ring would be the one that contrasted most with the background. Obviously a preference for black-and-yellow over black could the not be attributed to the greater contrast. However, it might be argued that to the eyes of a *Philanthus* our 'yellow' might possibly appear quite different from the colour of sand, and that it might even contrast more than black; alternatively it might be that, whatever the background colour of the soil, the wasps preferred yellow objects to black ones. We have tried to meet these possible objections by a number of control tests.

[1] The rings were 4·5 cm wide, hence the checks were appr. 1 sq. cm each. The results of this test raise the question as to whether this small pattern might not have been perceived as an undifferentiated area of uniform colour. The only fact known about the visual acuity of *Philanthus* is that they are able to perceive an object of the size of a honey bee from a distance of at least 30 cm. Since it is probable that a considerable number of decisions (though not all, cf. experiments 17 and 34) are made from a height of at least 25 cm, it is quite possible that some of the 'black' choices were attributable to the smallness of the checkerboard pattern.

Fig. 38. Experiment 7.

Fig. 39. Experiment 8.

Fig. 40. Experiment 9.

Experiment 7 Training to 6 black and 6 yellow cubes, each measuring 4 × 4 × 4 cm, arranged alternately in a circle (Fig. 38). (This switch to solid objects, which was done for reasons that need not concern us here, does of course not affect the validity of the experiments, since the objects differed in colour only).

Wasp No.	Black	Yellow
29	10	5
30	4	4
	14	9

Experiment 8 Training to a black versus a yellow ring, both three-dimensional (Fig. 39).

Wasp No.	Black	Yellow
31	8	3
32	4	7
	12	10

Thus if the wasps could be said to prefer either of the colours at all, it would seem that they had a preference for black rather than yellow. Because these experiments are so few in number, we did a few additional experiments, which were at the same time more straightforward.

In these subsequent tests we compared rings with bold black and yellow patterns with others where the pattern was more detailed.

Fig. 41. Experiment 10.

Since in all patterned rings one half of the surface was black and the other yellow, any effect of preference for one of the two colours was eliminated.

Experiment 9 A ring with 4 black and 4 yellow sectors versus one with 16 sectors of each colour (Fig. 40).

Wasp No.	Large pattern	Small pattern
33	3	4
34	5	7
35	5	4
	13	15

Experiment 10 One ring with 8 black and 8 yellow sectors versus one with 32 sectors of each colour (Fig. 41).

Wasp No.	Large pattern	Small pattern
36	0	6

Experiment 11 One ring with 8 black and 8 yellow sectors as above, versus a ring with small checks (32 sectors each subdivided concentrically) (Fig. 42).

Wasp No.	Large pattern	Small pattern
37	15	30
38	20	37
	35	67

Here, too, we find a preference for the landmark with the more detailed pattern. It is of course not surprising that the preference is not quite so clear-cut as when plain is pitted against patterned, for the differences between two types of landmarks used in the last 3 experiments were much less pronounced to us than those used in experiments 2–6.

The number of choices per wasp in experiments 9 and 10 was comparatively small; these observations date from the first summer, when we had not yet discovered that discrimination becomes more

Fig. 42. Experiment 11.

pronounced with longer training. We therefore consider experiment 11, which was done with two long-trained wasps in the last summer of our studies, the most reliable of all.

We argued further that if the choice of the patterned rings in preference to the black ones were a consequence of the 'yellow' colour *per se*, this preference should disappear, or even be reversed, if a plain yellow ring were substituted for the plain black ring.

By chance we found that, with other colours, such an overall preference did actually play a part. In 1937, we used white as the brighter of the two colours. When wasps trained to a black and a black–white patterned ring were given a preference test, they chose the patterned ring, but when they were then offered the patterned against a uniformly white ring they alighted more often inside the latter. Obviously, our 'white' was for some reason (U.V.?) a more effective colour. That this was not true of our 'yellow' was already apparent in experiments 7 and 8, and is shown once more by the next experiment.

Experiment 12 A number of wasps which had already been used in other experiments were tested once more on this particular point. This we did by offering them in the test proper rings with a black and yellow pattern versus plain yellow rings.

Wasp No.	Patterned	Yellow
15	6	0
183	3	3
184	8	2
185	4	0
	21	5

The overall conclusion must be that, although different colours may have different value as landmarks, the preference shown in the above series of experiments was due to patterning *per se*.

This preference for patterned landmarks appears however to be much weaker than that reported for the honeybee by Hertz (5). Admittedly differences in experimental conditions preclude a close comparison between the two species. For one thing, Hertz studied orientation in a different context (the finding of a food source), and

157

Fig. 43. Experiment 13.

for another she did not observe preferences after a prolonged period of training—rather she studied 'spontaneous' preferences (which of course must have been influenced by the bees' natural pre-test experiences). For practical reasons a detailed analysis of the kind carried out by Hertz on her bees was out of the question with *Philanthus*. We have to be satisfied with the statement that homing *Philanthus* females learn patterned landmarks more readily than uniformly coloured ones.

4. Three-dimensional and Flat Objects

Our next objective was to study the effect of solidity *per se*. Since solid objects cast shadows, and so are more patterned than flat objects even to animals that have no true depth perception, we had to reduce as much as possible the chances that any preference for solid landmarks could be ascribed to their shading. We therefore started by using three-dimensional rings and blocks with as simple contours as possible.

Experiment 13 Six wooden cubes 4 × 4 × 4 cm (painted dull black so as to reduce contrast between the various sides) versus 6 dull black flat 4 × 4 cm zinc squares (Fig. 43).

Wasp No.	Three-dimensional	Flat
39	5	0
40	7	0
41	4	0
42	1	0
	17	0

The same experiment was made with a female *Cerceris arenaria* L.

Three-dimensional	Flat
16	1

Experiment 14 Since cubes are of course slightly more patterned than flat discs, we offered in this experiment a choice between 6 dull black hemispheres (r = 2 cm) and 6 dull black flat circular discs of the same radius.

Fig. 44. Experiment 15.

Wasp No.	Three-dimensional	Flat
43	18	0
44	1	2
45	4	0
46	5	0
47	2	0
48	4	0
49	3	0
50	4	0
51	2	0
	43	2

Experiment 15 A three-dimensional ring versus a flat ring of the same colour (black, yellow, or black and yellow) (Fig. 44).

Wasp No.	Three-dimensional	Flat
52	8	0
53	8	0
54	6	0
55	12	0
56	8	0
57	14	0
	56	0

After we had thus shown conclusively that three-dimensional objects were greatly preferred as landmarks to flat objects, we felt that further analysis of this 'three-dimensional' property might be rewarding. The following potentially relevant characters were chosen for investigation. We thought that the greater attractiveness of the three-dimensional objects used in our tests so far might be attributable to either (1) their greater surface area; or (2) the casting of shadows; finally (3) we thought it conceivable that real stereoscopic vision might enable the wasps to perceive solidity as such, i.e. independent of shadow effects. We did not attempt to study any other factors (cf. Höber, 7), partly because the experiments reported below showed to our satisfaction which was the most important characteristic, and partly because our field conditions did not allow further work.

Experiment 16   In order to test the effect of surface area we trained the wasps with 6 dull black hemispheres of radius 2 cm, and 6 dull black discs of equal surface area, i.e. having a radius of $2\sqrt{2}$ cm

Wasp No.	Three-dimensional	Flat
58	15	1
59	13	1
60	12	2
61	8	0
	48	4

Experiment 17   We then argued that if the size of the surface area had any influence at all, we ought to be able to reverse the results by increasing the size of the flat discs. We therefore trained our wasps to 6 or 4 dull black hemispheres, as above, plus 6 or 4 dull black discs with a radius of 5 cm (Fig. 30; top right).

Wasp No.	Three-dimensional	Flat
62	21	14
63	38	3
64	4	2
65	14	0
	77	19

Even with this arrangement the three-dimensional objects had a surprisingly high score. We could therefore safely assume that the size of the surface area played at best a very minor role in the greater attractiveness of hemispheres.

Next we studied the role played by the shadows cast by the solid landmarks. The shadow projected on to a flat surface might of course act as two-dimensional patterning, and so account for the preference for solid objects. If that were so, plain-coloured, three-dimensional objects should lose some of their attractiveness if offered against flat objects patterned in two colours. The other possibility was that it was not just the breaking up of the surface, but the special form of the shadows which operated as the criterion of three-dimensional bodies. Both possibilities were tested.

Experiment 18   Six dull black hemispheres (r = 2 cm) versus 6 discs

160

Fig. 45. Experiment 18.

of the same radius bearing a black cross on a yellow background or (in the case of wasp 66) on a white one (Fig. 45).

Wasp No.	Three-dimensional	Flat
66	26	6
67	7	0
	33	6

Experiment 19 A dull black three-dimensional ring versus a flat ring with 16 black segments alternating with 16 yellow ones.

Wasp No.	Three-dimensional	Flat
68	10	0
69	6	0
70	12	0
	28	0

Experiment 20 As experiment 19, except that the three-dimensional black ring was replaced by a three-dimensional yellow ring

Wasp No.	Three-dimensional	Flat
71	10	0
72	8	0
	18	0

Thus the preference for three-dimensional objects was not diminished when the flat objects were broken up by patterns. The fact that wasp 66 chose the flat discs 6 times is presumably attributable to the preference for white reported above.

In order to test the influence of a direct shadow as a means of identifying a solid object, we painted the proper shadows on to a flat yellow ring (Fig. 46) and in:

Experiment 21 . . . compared this flat, shaded ring with a genuinely three-dimensional yellow ring.

161

THE ANIMAL IN ITS WORLD

Fig. 46. Experiment 20.

Wasp No.	Three-dimensional	Flat
73	18	0
74	29	1
75	15	2
	62	3

It appeared clear that this kind of shading had little effect, if any. Regrettably, we never compared a shaded with a non-shaded flat ring, but we thought at the time that we had found another, more direct method of testing the influence of shadow effects. That this method did not work was not discovered until the last summer of our joint researches was nearly over. We nevertheless consider it worth while to describe the experiments in question for they are of great interest in other respects.

When the wasps had been successfully trained to use hemispheres and discs, any preference for hemispheres could be largely eliminated by placing the discs upright into the sand in such a manner that one half was protruding above the soil surface and the tops of discs and hemispheres were of the same height (see experiment 27). This made us try the following experiment.

Experiment 22 After the wasps had been trained with 6 dull black hemispheres (r = 2 cm) and 6 dull black discs of the same radius, we placed 6 discs upright around one of the sham nests so that they were parallel with the rays of the sun and thus cast hardly any shadow. Around the other, we also placed upright discs, but this time we arranged them at right angles to the sun's rays (Fig. 47). In this manner we contrasted maximum and minimum shadow. This experiment was carried out only in bright sunshine.

Fig. 47. Maximum shadow (left) against minimum shadow.

Wasp No.	Maximum shadow	Minimum shadow
47	17	7
49	3	0
51	3	9
	23	16

The experiment with wasp 51 was made in the last summer and only then did we notice the interference caused by certain factors which we had not recognised before. We did our tests at appr. 1 p.m.; there was usually a slight, sometimes a moderate, westerly wind, i.e. the wind blew in a direction at right angles to that of the sun's rays. The wasp, arriving as usual against the wind, would therefore see the discs which were standing parallel to the sun's rays broadside on but see the others edge on. It seemed likely that this had affected their choice (cf. experiment 34). When we tested wasp 47, we had not been aware of this possibility, but as we always recorded the time of the experiment and the prevailing wind, we were able to check whether on that day the situation had been different (wind parallel to the rays of the sun). This had indeed been the case: our records showed: time, 2 p.m., wind NNE. But though our experiments cannot be used as evidence on the point we sought to elucidate, they nevertheless suggest that if circumstances like those described can eliminate it completely, the influence of shadow must be very small.

Another possibility was that the deeper black of the shaded sides of the hemisphere distinguished them for the wasps from the flat discs which received direct sunlight and therefore appeared lighter. If this were so, it could be expected that this preference would be reduced if the hemisphere were replaced, *in the tests*, by hemispheres of a lighter colour. That this had no effect became clear from our next experiment.

Experiment 23 After the wasps had been trained to 6 dull black hemispheres (r = 2 cm) and 6 dull black discs of the same radius, the black hemispheres were replaced in the choice tests by white, yellow, or light grey hemispheres.

Wasp No.	Three-dimensional	Flat
47 (yellow)	8	0
49 (grey)	4	0
51 (white)	8	0
	20	0

163

Finally we shaded the whole arrangement of flat and three-dimensional landmarks from direct light, during the test, so that all strong hard shadows disappeared. Of course there remained some unevenness but the illumination was much more diffuse and very different from that in bright sunlight.

Experiment 24   After training (this time to dull black pine cones and a dull black flat ring—experiment 1) we offered a dull black three-dimensional ring and a dull black flat ring. Even in this choice situation the wasps invariably chose the three-dimensional ring. The same was true when we shaded the two sham nests and the surrounding area (appr. 80 × 100 cm).

Wasp No.	Three-dimensional	Flat
6	5	0
8	21	0
	26	0

Thus we have no evidence at all that the preference for three-dimensional landmarks is associated with shadows, though no evidence either that shadows may not have some slight influence. However, I believe we are justified in saying that the influence of shadow, if any, must be very small indeed.

Since our tests had shown that neither a fragmentation into lighter and darker surfaces, nor the effect of shadows could explain the preference for three-dimensional objects, we had to look for other factors. We tried to find out to what extent the landmarks were in fact perceived three-dimensionally. To this end we tested our wasps not as hitherto on flat discs and objects which stood on the surface of the soil, but on flat discs and objects which formed a depression in the soil. For technical reasons we used not hemispheres but hollow cones, the bases of which were of the same diameter as the discs. These cones were buried until their rims were level with the soil, and did not show above it at all.

Experiment 25   Six hollow paper cones (of the shade 'Hering grey 22') (r = 2 cm) buried up to their rims and 6 discs (r = 2 cm) cut from the same paper (Fig. 48; top left).

Wasp No.	Three-dimensional	Flat
76	5	7
77	0	2
78	11	10
	16	19

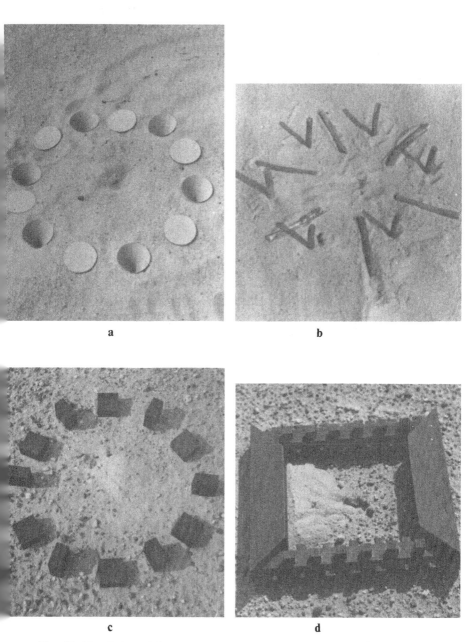

Fig. 48. Top left: training to 'sunk' cones and flat discs. Top right: training to standing and horizontal sticks. Bottom left: training to standing and flat half-cubes. Bottom right: training to smooth and

Training with this arrangement, like that with flat objects, was difficult. There was no question of any preference for the sunken cones; as far as it is possible to draw any conclusions at all from such meagre data, rather the opposite might be inferred. The behaviour of the wasps suggested, however, that they were aware of the difference between the two types of objects: they often flew around them for a considerable time, and not infrequently would alight inside one of the cones.

The following summer we supplemented these experiments with a new series, in which we contrasted hollow cones buried up to their rims with cones standing on the ground. The actual shapes of the two types of cones, now made of tinplate and painted white, grey or black, were identical. To human eyes the inverted, buried cones looked much less 'three-dimensional' than those standing on the ground. But of course we too depend for our perception of three-dimensional shapes on more than binocular vision alone.

Experiment 26 Six or 4 dull black, dull white, or dull grey 'upright' cones versus 6 or 4 cones of the same colour, buried upside-down.

Wasp No.	Standing	Buried
79	6	4
80	2	2
81	5	6
82	6	0
83	11	0
84	4	0
85	38	5
86	21	2
87	4	0
88	11	2
	108	21

Again there was a strong preference for the upright cones, although it was less pronounced than might have been expected from experiment 25. Yet we had now to admit that our first conclusion, namely, that three-dimensional objects are preferred to flat objects, had been put in too general terms: even though the relatively high score for 'buried' might suggest that three-dimensional objects which did not protrude above the surface of the soil (i.e. 'sunk' cones) were perhaps slightly preferred to flat objects, they certainly were much less favoured than objects standing on the ground and rising above it. Subsequent experiments were designed to test this point further.

Fig. 49. Discs facing entrance versus hemispheres.

Experiment 27 After training the wasps with 6 dull black hemispheres (radius 2 cm) and 6 dull black discs (same radius) we compared not, as before, hemispheres with flat discs, but hemispheres with vertical discs, half buried and half protruding from the sand. In some of the tests these discs were all arranged in one orientation as in experiment 22—either all parallel to the sun, or all at right angles to it (minimum versus maximum shadow)—in others as a 'coronet', their broad sides turned towards the entrance to the burrow at the centre (Fig. 49); and yet in others as a 'star', their narrow sides pointing to the entrance (Fig. 50). Results of these 4 arrangements were as follows:

Wasp No.	Discs at right angles to the sun's rays	Hemispheres
43	3	6

Wasp No.	Parallel to the sun's rays	Hemispheres
43	4	10
45	6	6
46	0	6
	10	22

Wasp No.	'Coronet'	Hemispheres
43	3	2
45	4	2
46	1	1
	8	5

Wasp No.	Star	Hemispheres
45	0	4
47	0	2
	0	6

Fig. 50. Discs radially arranged versus hemispheres.

Taken together these figures suggest that the upright discs score only slightly lower than the hemispheres, despite their different shape. As we shall discuss later, the 'coronet' arrangement is actually the 'strongest' one, the star arrangement the 'weakest'.

We made further progress when, almost accidentally, we tried a similar experiment using cubes versus flat squares, as in experiment 13.

Experiment 28 After having trained our wasps with 6 dull black cubes, 4 × 4 × 4 cm, and 6 dull black flat squares, 4 × 4 cm, we compared cubes to squares inserted upright into the sand. To our surprise, the result was as follows:

Wasp No.	Upright squares	Cubes
41	0	15
89	0	4
39	0	5
	0	24

The main difference between this test and experiment 27 appeared to be that in experiment 27 the hemispheres and the flat upright objects were of exactly the same height, while in experiment 28 the upright but flat objects, the squares, were appr. 1 cm lower than the cubes, because they had had to be stuck into the sand appr. 1 cm deep. We therefore tested this factor 'elevation' on a few wasps, some of which had been used before, but were now retrained.

Experiment 29 The training was the same, but now we pressed the cubes 1–2 cm deep into the sand, with the result that their tops were now 3–5 mm *lower* than the top edges of squares standing upright.

Wasp No.	Squares standing upright	Cubes pressed down
39	6	9
89	7	7
90	8	5
	21	21

This very clear result suggests strongly that the height of the object

168

above the ground is the decisive factor. Even comparatively small differences in height appear to have an effect; and at the same time, the last experiment made it clear once more that the surface area of the landmarks is relatively unimportant. We then tried to test this influence of height by other, more direct methods. We trained the wasps with 12 half-cubes $4 \times 4 \times 2$ cm, one half of which we stood upright, while the others were laid flat on their 4×4 cm side. We knew by then that the surfaces facing the burrow were of importance (see experiment 34 below) and we therefore arranged all half-cubes in such a manner that the surface turned towards the nest was 4×2 cm.

Experiment 30 Six dull black half-cubes ($4 \times 4 \times 2$ cm) standing upright, versus 6 lying on the ground (Fig. 48; bottom left). As a rule we applied this test after training the wasps with these two landmarks but occasionally after they had been trained with 6 cubes and 6 flat squares. In the latter case we wanted to find out which property of the cube is the most important; its height or its width. The small numbers preclude separation on the basis of different training regimes.

Wasp No.	Upright	Lying on side
91	2	1
92	26	12
93	2	1
	30	14

The following summer we repeated the test in a more sophisticated form. As a result of other experiments (experiments 44 and 45), we thought it possible that the upright half-cubes were preferred because of their greater surface area (their invisible base was, of course, 4×2 cm, and that of the flat-lying half-cubes 4×4 cm). We therefore used 'quarter-cubes', $4 \times 2 \times 2$ cm, standing upright, and half-cubes, $4 \times 4 \times 2$ cm, lying flat. The surface turned towards the nest entrance was 2×4 cm for both.

Experiment 31 Six dull black upright $4 \times 2 \times 2$ cm quarter-cubes versus 6 flat lying $4 \times 4 \times 2$ cm dull black half-cubes. This test was usually given after training with the two markers in question, but occasionally after training to wooden cubes and flat squares.

169

Wasp No.	Upright	On side
94	4	4
95	21	12
96	9	5
97	7	11
98	104	49
99	4	5
100	6	15
101	17	13
102	6	6
103	37	4
104	6	2
	221	125

There was thus not much change in the degree of preference for 'height' over 'width'. However, it is interesting to consider the case of wasps No. 97 and No. 100, which apparently showed a preference for 'width' rather than 'height'. In this context we must remember that, strictly speaking, our experiment offered two different factors in competition: on the one hand, height from the ground (which would favour the 'uprights') and on the other, total size of surface and of surface projection (which was largest in the other landmarks). It is conceivable that different individuals attached different weight to these factors in their choices. Another aspect of this experiment will be discussed later, namely that wasps that had not been tested often (Nos 94, 99, 102, 104) showed less definite preferences than wasps that had been tested a great many times (Nos 95, 98, 103).

The same experiment was done with the *Cerceris arenaria* specimen mentioned under experiment 13. Its performance was most convincing: 16 to 'upright' versus 1 to 'on side'.

In this connection we should also like to mention:

Experiment 32 Twelve round sticks (r = 6 mm, length 12 cm) were placed in a circle round the nest, an upright stick alternating with one lying on the ground and pointing towards the nest (Fig. 48; top right). In the actual tests we did not change the actual position of the sticks but merely changed each upright stick to one lying flat (again arranged in a 'star' pattern) and vice versa. We had to take this precaution because the sticks were rather roughly prepared and not absolutely identical; in this way we eliminated the effect of individual differences between them.

170

Wasp No.	Upright	Flat
105	17	2

Since, viewed from the nest, the upright sticks in this 'star' arrangement presented a much larger surface area than the sticks on the ground, we next did:

Experiment 33   Here the sticks on the ground were laid out in a 'coronet' arrangement.

Wasp No.	Upright	Flat
105	6	0

The results of all these experiments taken together leave no doubt: the three-dimensional landmarks are preferred chiefly because of their height above the ground. In comparison, surface area, patterning and shading appear to be of little importance.

In this context we wish to describe one more experiment which throws light on the important question of the exact moment at which the wasp finally 'decides' on which landmark to go by.

Experiment 34   After the wasps had been trained with 6 dull black hemispheres and 6 dull black circular discs (experiment 14) we offered 6 discs standing upright, half buried in the sand, in the 'coronet' arrangement, i.e. presenting their broad sides to the centre, against 6 discs, likewise upright, but in the 'star' position, i.e. presenting their narrow sides to the centre.

Wasp No.	Coronet	Star
46	4	0
106	34	12
	38	12

It was most interesting to see the wasps make their choice. Frequently a wasp would come down to within 1 cm of the ground at the 'star' and would hover for a second but would not actually alight. Suddenly it would fly over to the coronet and settle there. Only in a small proportion of the star choices did the wasp make for the centre (i.e. the nest entrance); in the majority of cases it settled more or less off-centre, a somewhat unusual behaviour, for in general the wasps were meticulous in choosing the right position in relation to the landmarks, even when we had failed to make the entrance to the sham nest at exactly the right place (we had found that we could afford to be rather casual in the preparation of the sham nest without affecting the orientation of the wasps).

This choice of an off-centre position and the rest of the behaviour observed in experiment 34 suggests, we believe, that the 'star' discs are almost invisible to the wasp from the centre position and that this made them either stop at the last moment or alight on a spot where it could see at least some discs broadside on.

These observations make it very clear that the wasp scrutinises the landmarks in the immediate surroundings of the nest up to the very last moment until actually alighting. The importance of the sand patch and other distinctive aspects of the burrow itself are negligible compared to that of the landmarks. And herein we may, perhaps, find at least a partial explanation of the fact that height is the most important characteristic of solid objects: flat objects simply cannot be seen from the entrance of the burrow. But this is certainly not the whole explanation. We always tried to pin-point the moment of decision, and it was not often as late as in experiment 34. In fact, the majority of decisions between 'high' and 'flat' were made from a considerable height, i.e. generally before the wasp had come down to appr. 25 cm from the ground. In experiment 17, in which we offered very large flat discs and small hemispheres, we expected that the wasps, homing from a great height, would at first fly towards the large discs and only later, after having descended to a much lower level, veer towards the smaller hemispheres. But careful observation of the 73 cases in which a wasp chose the hemispheres (cf. experiment 17) did not bear this out. Even when faced with the very large discs the wasps chose the hemispheres from a considerable distance: in most cases, as stated above, they did so before coming within 25 cm of the soil surface.

We did not make further attempts to find out how the wasps recognised solid objects, e.g. whether they were guided by true binocular vision, since preliminary experiments met with difficulties that could not be overcome in the field.

5. Three-dimensional Patterning

Our first experiments to test the effect of three-dimensional patterning suffered from the fact that we underestimated the influence of even minor differences in height between different landmarks. The cubes already described, or three-dimensional rings served as landmarks with a smooth surface; for patterned landmarks we used the same cubes or rings but with their surface covered with a number of little cube-shaped knobs, each measuring $1 \times 1 \times 1$ cm. In our original experiments these 'knobbly' markers were greatly preferred by wasps which had been trained with the basic smooth forms. However, after

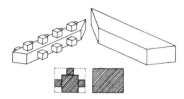

Fig. 51. Experiment 35: objects used; above, pieces of frames; below, cross-sections.

a series of control tests with smooth but slightly larger landmarks, we thought it likely that this preference could be attributed to the difference in overall height of the two groups of landmarks. In the end we settled on the following: for a smooth marker, we used a wooden frame made of 4 oblong pieces, joined at the ends; each piece measured 4 × 3 cm in cross-section, 20 or 30 cm at the outer edge, and 12 or 22 cm at the inner edge; as the patterned counterpart we used a similar square, with the same outer dimensions, only this time the cross-section of the pieces was made 2 × 2 cm and the inner edges 16 or 26 cm; onto this frame we had stuck a number of our 1 cm cubes: 7 or 10 on the outer, 6 or 8 on the inner, and 6 or 9 on the top surface of each side (Fig. 51). All objects were painted a dull black. We had prepared two sizes of pieces because we intended to operate with an outer and an inner frame, as we had done with rings. But we soon discovered that such an arrangement made the wasps consistently choose the type of frame that had been on the inside during training. Thereafter we constructed our training frames from both types of pieces: In some tests the training arrangement was smooth front and back, and patterned right and left; in others, the smooth pieces were at the sides and the patterned pieces front and back (Fig. 48; bottom right). During the test itself we offered two frames, one made entirely from smooth, the other from patterned pieces (Fig. 52).

Experiment 35 A frame made of dull black, smooth pieces and another made of dull black patterned pieces (Fig. 52).

There was thus a clear preference for the patterned frame. We are now in a position to arrange our landmarks in a descending order of preference, a hierarchy of landmarks, as it were:

Three-dimensional with 'knobs' > three-dimensional with smooth unbroken surface > flat with two-dimensional patterning > flat and uniform in colour.

Following these experiments, in which only one of the above characters had been investigated at any one time, we trained some wasps to the two extreme ends of our series and tried to ascertain

173

Wasp No.	Smooth	Patterned
107	0	9
108	0	18
109	0	14
110	4	5
111	2	5
112	4	14
113	4	10
114	8	7
115	1	20
	23	102

whether all these features really did affect their choice. After training was completed we would make sure that one of the two types was preferred, and then offer an object which differed from the preferred type in no more than one of its characteristics.

We used two different arrangements:

(1) Training with a three-dimensional ring which had a flat surface pattern and with a flat ring coloured plain black:

Experiment 36 Flat ring patterned like the three-dimensional training ring versus flat black ring.

Wasp No.	Patterned	Black
116	5	1
117	5	0
	10	1

Experiment 37 A three-dimensional ring versus a flat ring of the same colour.

Wasp No.	Three-dimensional	Flat
118	2	0
119	1	0
120	3	0
	6	0

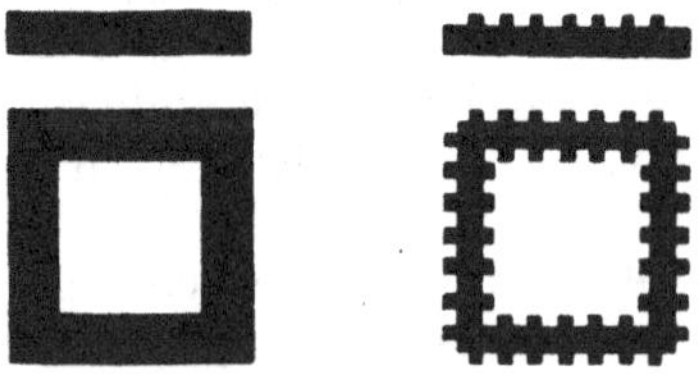

Fig. 52. Experiment 35.

174

(2) Training with dull black pine cones and a smooth, dull black, flat ring, as in the first experiment, then testing with intermediate objects:

Experiment 38 Three-dimensional ring against a flat ring of the same colour.

Wasp No.	Three-dimensional	Flat
2	4	0
3	10	0
	14	0

This demonstrated the effect of three-dimensionality alone.

Experiment 39 Three-dimensional, smooth black ring versus dull black pine cones.

Wasp No.	Three-dimensional smooth surface	Cones
2	1	6
3	0	4
4	0	4
7	0	6
9	0	6
	1	26

This showed that the smooth ring was clearly distinguished from a ring of cones.

The effect of three-dimensional patterning of outline was once more demonstrated in two different ways:

Experiment 40 The pine cones used in training versus wooden cubes 4 × 4 × 4 cm in size, of the same colour.

Wasp No.	Cones	Cubes
4	16	5
5	2	2
6	2	2
	20	9

The greater height of the cubes no doubt explains (though not completely) why the preference for the pine cones is not stronger—the latter were of course three-dimensionally patterned.

Experiment 41 The same pine cones used in the training versus the same wooden cubes as in experiment 40, but this time the wooden cubes were placed close together so that there was no space between them as in experiment 40.

Wasp No.	Cones	Cubes closed up
5	5	0
6	6	0
	11	0

As soon as the gaps between the individual cubes had disappeared, the preference for the cones became very strong, notwithstanding the fact that the cubes were put nearer to the nest (this was a consequence of their being placed close together) and that they were higher than the cones.

That three-dimensional patterning of the surface is not only effective as such, but acts also as flat patterning, i.e. in projection on a flat surface, was shown by:

Experiment 42 A three-dimensional, smooth, black ring versus a three-dimensional ring patterned in yellow and black.

Wasp No.	Cones	Patterned ring
9	6	0

Hence we may say that pine cones differ from flat rings by being higher, by being three-dimensionally patterned, and by their pattern of light and shade seen in projection.

6. Size

Experiment 44 Four large wooden cubes, measuring $8 \times 8 \times 8$ cm and 4 small cubes measuring $4 \times 4 \times 4$ cm were so arranged, during training and during the test, that the surfaces facing the burrow of the real nest or the sham nest respectively were equidistant from the entry hole (Fig. 53).

Wasp No.	Large	Small
121	6	0
122	9	1
	15	1

176

Fig. 53. Experiment 44: large and small cubes.

Since, as we have seen, height is a very important factor, we changed the test slightly by slicing the large cubes in half, so that they were of the same height as the small ones.

Experiment 45 Four half-cubes measuring $8 \times 8 \times 4$ cm standing on a square side versus 4 small cubes $4 \times 4 \times 4$ cm, again placed at the same distance from the burrow entrance.

Wasp No.	Large	Small
123	7	0
124	5	0
	12	0

The larger objects were clearly preferred. This preference explains why, in experiment 17, the large flat discs were chosen comparatively frequently. It also probably accounts, at least partly, for the relatively low preference for the upright quarter-cubes found in experiment 31. The half-cubes lying on their square sides presented, when seen from above, 4 times the surface of the upright quarter-cubes.

7. Brightness Contrast

As early as 1934 and 1935, when we studied the role of the sand ejected from the burrow as a landmark, we had discovered that the contrast between the sand patch and the ground was of great importance. On open stretches, even of bare sand, these patches of sand were to us by far the most conspicuous signs. But, surprisingly, the wasps themselves would always follow such things as a tuft of grass, even if we moved it as much as 50 cm from a nest which still had its patch of sand intact. This led us to investigate more closely the role of the sand patch as a landmark for orientation.

It soon became clear that its value as a landmark depended on the colour of the surrounding surface. This varied widely from bare sand to, in many places, a closed mat of dark brown mosses and lichens.

In some places the ground was so closely covered with burrows and their sand patches, that a wasp might well steer by neighbouring sand

177

patches as well as its own. This suggested the following series of experiments.

Experiment 46 We chose a nest on a sandy site. As always on such sites, the patch of sand was slightly more yellow than the surface sand but did not really contrast very strongly with it. While the wasp was out hunting we gently blew the loose sand away until the ground underneath reappeared, but were careful not to touch the entrance of the burrow. Twenty cm further on (at right angles to the direction of the wind, and therefore also at right angles to the flight path of the wasp) we fashioned a new sand patch, trying to make it as similar to the original patch as possible. After the homing wasp had made a number of choices, the artificial sand patch was, in turn, blown away and the original patch restored. This experiment was done with 7 wasps. In the table, the location of the sand patch is indicated by italics at the head of the column.

Wasp No.	Test		Control	
	nest	*sham nest*	*nest*	sham nest
125	4	6	6	0
126	1	4	8	2
127	0	5	5	0
128	8	0	5	0
129	6	6	1	0
130	5	0	—	—
131	7	3	4	0
132	5	0	5	0
	36	24	34	2

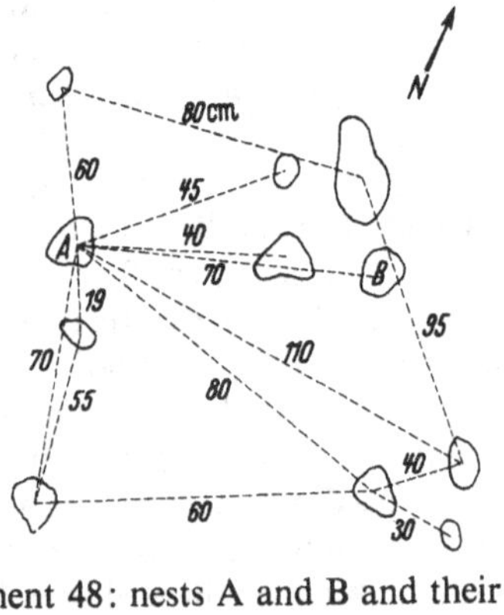

Fig. 54. Experiment 48: nests A and B and their neighbours.

178

We could already see that the sand patch was not without its significance even where it was rather similar in colour to the ground, though its value as a landmark was not very great.

Experiment 47 This followed the same lines as experiment 46, but this time the nest was in a colony on a dark, brownish-black background (largely covered with dry lichens) (Fig. 30; bottom left).

While 1 of the wasps took absolutely no notice of the sand patch, the other 5 appeared to rely on it completely.

Wasp No.	Test		Control	
	nest	*sham nest*	*nest*	sham nest
133	5	0	5	0
134	0	5	5	0
135	0	5	5	0
136	0	5	5	0
137	0	5	5	0
138	0	5	5	0
	5	25	30	0

Experiment 48 During the absence of wasps 139 and 140, each of which had built its nest (A and B respectively) on a dark background among many others (Fig. 54) the sand patches from nine neighbouring nests (but not those of nests A and B) were carefully obliterated, and new sand patches were made 20 cm to the NE of each of these nests. A and B were given two 'sham nests', one (S 1) 20 cm to the NE, one (S 2) 20 cm to the SW of the original nests.

When wasp 139 (A) returned, it chose:

S (A) 2	S (A) 1
0	5

Next, we once more obliterated the sand patches belonging to the neighbouring nests and faked new ones, this time 20 cm to the SW of all nests.

This time wasp 139 chose:

S (A) 2	S (A) 1
1	0

At that moment wasp 140 of nest B came back and chose:

Wasp No.	S (B) 2	S (B) 1
140	4	0

A moment later another 'B' wasp returned; as we found out later, there were two wasps that had their nest at B and they had jointly fashioned the sand patch at that point. This second 'B' wasp chose:

Wasp No.	S (B) 2	S (B) 1
141	2	0

We inadvertently confused these two wasps and could no longer say which was No. 140 and which No. 141. But this of course in no way invalidated our results.

Wasp No.	S (B) 2	S (B) 1
140 or 141	3	0

For confirmation, the sand patches of all the neighbouring nests were now moved again to 20 cm NE of their original position (i.e. 40 cm to the NE from where they had just been).

Wasp No.	S (B) 2	S (B) 1
140 or 141	0	3

In spite of the small figures of each test, the total score (18 against 0) made it abundantly clear that these wasps were guided predominantly by the position of neighbouring nests; the fact that their own sand patches had remained completely untouched did not appear to matter to them at all.

We repeated this experiment later with another group of nests on a similar dark background. The result was the same, and we can dispense with further records.

Experiment 49 We repeated the experiment with 3 groups of nests and 5 wasps (Nos 142, 143, 144, 145 and 146). The nests were spaced similarly to the 2 nests of experiment 48, but this time the background was light (sand). The result is quickly told: All our careful shifting of sand patches and fashioning of sham nests was of no avail. Not once, in 63 choice situations, did we succeed in leading the wasps astray; they invariably chose their real nests. Thus in this case there was absolutely no sign of neighbouring nests being used as landmarks.

Clearly sand patches in front of the wasp's own burrow and those belonging to neighbouring nests have greater value as landmarks on a dark than on a light background. This leads one to assume that the contrast between sand patch and background must be the decisive factor. But though we started a number of experiments designed to corroborate this conclusion, adverse weather conditions prevented their completion.

8. The Distance from the Nest

When discussing our experiments on three-dimensional patterning we mentioned that with our original arrangement (two frames, one inside the other), a preference for patterned surfaces could not be demonstrated because the influence of the patterning was over-shadowed by the distance of the different markers from the next. The experiments in question are reported below:

Experiment 50 The wasps were trained with a small frame made of smooth pieces, dull black in colour, and a larger frame made of dull black but patterned pieces. The pieces used were the same as described in experiment 35.

Wasp No.	Smooth (inside)	Structured (outside)
147	10	4
148	18	0
149	32	1
	60	5

Experiment 51 As experiment 50, but the position of the two types of markers *during training* was reversed.

Wasp No.	Smooth (outside)	Structured (inside)
150	8	8
151	0	12
152	3	21
	11	41

While these two experiments did not teach us anything about the value of patterning they did show that the wasps had responded mainly to the inner of the 2 frames, i.e. the only frame that was visible when the wasp was sitting in or near the entry to its burrow.

In those experiments where even a wasp on the ground could see both landmarks, that is to say in many of our experiments with rings, no such strong preference for the inner ring was observed, and so the factor 'nearness' had never given us any trouble. But now we thought we should isolate the effect of this particular factor by offering sets of identical objects at different distances from the nest, but equally visible to a wasp even when it was sitting at its nest entrance.

181

Experiment 52 Four sand-coloured cubes, $4 \times 4 \times 4$ cm in size, placed 12 cm from the nest, and 4 identical cubes placed at a distance of 24 cm, all with a side facing the entrance (Fig. 55).

Wasp No.	Near	Far
153	5	0
154	15	0
155	8	3
	28	3

Although we were aware of the importance of what the wasp could see from the entrance to the burrow (cf. experiment 34), we still needed to find out whether or not the preference shown for landmarks closer to the nest was a preference for higher or, more generally, larger objects. Of course for a wasp sitting at the entrance to its burrow the visual angle subtended by the nearer cubes is twice as great as that subtended by the other set.

Experiment 53 The same arrangement as in experiment 52, but this time the more distant cubes measured $8 \times 8 \times 8$ cm.

Wasp No.	Near	Far
156	0	6
157	2	7
	2	13

Rather surprisingly, the more distant but larger cubes are now preferred. Considering that from the nest entrance both types of landmark would be subtended by the same angle this seemed to suggest that the final decision is made before the wasp touches down. It also pointed for the first time to the possibility that homing wasps use distant landmarks even on their last lap. We ourselves did not pursue this further, but G. van Beusekom will shortly report on a study he has made of this. However, where landmarks are of the same size, our findings indicate that those nearest to the nest are the most important for orientation.

9. When Learning Occurs: the Function of 'Locality Studies'

The idea that the time of exposure to objects affected their use as landmarks occurred to us when we tried to find out more about the wasps' 'locality studies' (a term coined by Rau which has a built-in functional interpretation). For this, we tried to train a wasp on her

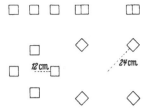

Fig. 55. Experiment 52.

outward flight to use a circle of 20 pine cones. We placed the pine cones round the nest while the wasp was inside, and recorded the duration and form of her departure flights. As usual, we rearranged the cones around a sham nest during the wasp's absence, and recorded her choices. In these experiments we limited our observations to wasps which made one uninterrupted departure flight, rising ever higher and circling ever wider without interrupting their flight by alighting on or close to the nest, for we wanted to eliminate the effect of such interim returns. In order to have the most favourable conditions possible, all these experiments were done in colonies where the background was sand-coloured, so that our pine cones stood out clearly.

Experiment 54 Training with a circle of pine cones during one locality study; testing by shifting the circle before the wasp had returned. In the tables showing the results of these experiments, the location of the landmarks is once more indicated in italics.

Wasp No.	Test		Control	
	nest	*sham nest*	*nest*	sham nest
158	1	0	—	—
159	3	0	4	0
160	4	0	2	0
161	5	0	5	0
162	5	0	5	0
	18	0	16	0

There is not the slightest trace of learning here, and so it seemed that one locality study was not sufficient to train a wasp to what had been proved to be 'powerful' landmarks.

Experiment 55 We therefore continued the training for a longer period. After placing the landmarks in position, we allowed the

183

wasps to fly away, come back, and fly away again. At the second homing we shifted the cones exactly as we had done in the previous series. This time we did not limit ourselves to wasps which had not made any landings in the middle of their flights, but used all the animals which had been trained.

Wasp No.	Test		Control	
	nest	*sham nest*	*nest*	sham nest
163	5	0	—	—
164	0	5	5	0
165	1	10	10	0
166	5	0	5	0
	11	15	20	0

Here, too, there was little evidence of learning. These results were rather unexpected, and we naturally asked ourselves by what method the wasps managed to find the correct spot when they returned from their very first flight away from the nest.

There was one circumstance which we thought might have a bearing on this problem: all the wasps tested by us had already made a great many flights to and from the nest before we started adding artificial landmarks to their environment, and must therefore already have learnt other landmarks. We reasoned that training is likely to be successful more quickly in a 'new' than in an already familiar environment.

A 'new' environment may come into being in two ways: (a) when a wasp digs a new burrow (the wasps, as we know, dig several burrows on various sites in the course of a season—cf. Tinbergen, **15**); or (b) when a long period of rain compels a wasp to stay 'indoors' for days on end, and the rain in the meantime completely alters the nature of the environment. In that case a wasp would *have* to learn the new situation even if she had not simply forgotten her previous training.

For various reasons which cannot be discussed here, it was not possible to experiment with new landmarks as soon as a wasp started building its nest. We therefore had to be satisfied with observing training effects during a wasp's very first locality study following a period of rain. The method was the same as in experiments 54 and 55.

Experiment 56 On the first sunny morning after a rainy period of appr. 2 weeks we started our observations on a group of nests on a light background before a single wasp had emerged. As soon as a wasp showed itself at the entrance to its burrow, we made it with-draw by a quick motion of the hand. Then we quickly arranged a circle of pine cones around its nest, and waited. After a while the wasp would reappear, would make one or more locality studies (the duration of which was carefully recorded) and would then fly off. At once, the circle of pine cones was moved to a sham nest. The wasp's choices upon its first return were recorded.

Wasp No.	Test		Control	
	nest	*sham nest*	*nest*	sham nest
167	2	0	—	—
168	0	5	5	0
169	5	0	—	—
170	0	6	5	0
171	0	5	5	0
172	0	3	4	0
173	2	6	5	0
174	0	10	2	0
175	0	18	8	0
176	8	4	0	3
177	1	4	1	0
178	2	0	—	—
179	0	12	1	0
	20	73	36	3

These results showed that most wasps which had made only one flight had learnt more than wasps 158–162 (experiment 54) and even considerably more than wasps 163–166 (experiment 53) which had all been tested after 2 departures and 1 return.

Some individual records support the conclusion that the 'locality study' is indeed what the term implies and serves to help the wasp memorise the position of the nest in relation to landmarks in the area surrounding it. Wasps No. 174 and 177 made just one locality study lasting 13 sec; wasp 179, which was perfectly trained, was even more remarkable: its departure flight had taken only 6 sec!

Although the flight pattern called 'locality study' or 'orientation flight' has often been described for a variety of Hymenoptera, its actual function was implied rather than demonstrated until Opfinger (**14**) studied its significance in the honeybee. Here too it is used for rapid learning of landmarks. The facts that the functional context (departure from a newly found food source) was different from ours (departure from the burrow) and that our methods were different prevent a meaningful comparison of the results. Opfinger's bees concentrated less on nearby landmarks than did our wasps, but in both species the locality study has now been demonstrated to deserve this name fully; it is an extreme example of a sharply defined 'teachable' condition—a sensitive period in which a remarkable feat of conditioning occurs and, as was by now abundantly clear, conditioning which is highly selective as well.

Most of the experiments just reported date from 1934. In 1936 we once more tried to train 3 wasps (Nos 180, 181 and 182) to a circle of pine cones after a single departure from the nest, in the middle of a period of fine weather, a feat we had been unable to accomplish on earlier occasions. To our great surprise all three wasps concerned in that new experiment were perfectly trained. We attribute this result to the following circumstances: In the first place, all the nests which we had used for our first set of experiments in 1934 had been in a lichen-covered area dotted with isolated pine trees, whereas the three nests of 1936 were on a perfectly bare and open stretch of sand. That is to say that whereas in 1934 the nearest pines had been no more than 10 m away from our nests, the distance to the nearest pines in our 1936 experiment was over 60 m. In 1936 therefore the pine cones were almost the only cues available.

Secondly we found that the 1936 wasps made far more and far longer locality studies than the 1934 wasps had done, or indeed other wasps familiar with their environment. This, we presume, was because in this terrain devoid of landmarks, orientation was so much more difficult.

The locality studies of wasp 180 lasted 80, 40, 58 sec; those of wasp 181: 3, 100, 2, 2, 30, 6 and 12 sec; and those of wasp 182: 40, 36, 7, 15, 38, 21 and 12 sec.

This behaviour suggests that wasps nesting in bare featureless terrain adjust the number and the duration of their locality studies accordingly, and take care to memorise striking new landmarks immediately. E. Wolf (18) reports that honeybees show the same increase in 'reconnoitring' in terrain poor in effective landmarks.

On the basis of these experiments, we are, I believe, justified in assuming that in addition to characteristics of the landmarks themselves, other circumstances also determine whether a given object near the burrow is accorded a major or a minor role as a landmark. The first time a wasp surveys the locality of the nest, it acquires a knowledge of the relative position of existing landmarks in a very short time; it takes longer for it to take in landmarks that are added later, even if these landmarks are themselves 'strong' by reason of their shape and colour. There is a critical period in which the ability to memorise landmarks is at its height.

10. The Duration of Training: Progressive Elaboration of the Field of Perception; Abrupt Changes in the Choice of Landmarks; Discarding of Landmarks

The experiments which we are now going to discuss were not done specifically with the above problems in mind. They have all been reported previously but are now considered in a different context. While conducting these experiments, we noticed again and again that the perceptual achievements of our wasps changed during training; sometimes even during one series of trials.

As explained before we always began our tests by moving the entire training complex as a whole; before we started our choice experiments, we wanted to be sure that they used the whole complex of markers for orientation. Only when we were satisfied that they followed our land marks did we declare them ready for the preference test.

What happened in a great number of cases was as follows: A wasp had consistently alighted within the circle of landmarks when we moved the entire set. But when we separated the two components all the learning suddenly seemed to have disappeared; the wasp would choose neither of the sham nests but went to her own nest. We had this experience any number of times, for (always impatient to obtain our data before the weather broke) we tested each home-coming wasp to see whether it was ready for a preference test. But when we repeated the experiment after further training, it was as if the wasp had completely forgotten her original nest, for she now chose one or both of the two dummy nests. Part of the explanation seems to be that the natural landmarks used by the wasps before we laid out our landmarks gradually lost their potency as the animals became used to the new marks, until finally the former were replaced by the latter. Because we normally took care to use 'strong' landmarks, we usually witnessed a shift from weaker to stronger landmarks. But we were also able to observe the gradual nature of the switch when we attempted to train wasps to flat marks. It would sometimes take up to two whole days before a wasp could be made to follow such marks reasonably well. And since we were checking the progress of a wasp's training each time it returned from a flight, the gradual nature of the switch was confirmed time after time.

As will be shown later, there is a clear difference between this gradual shift and the sudden, abrupt changes in the hierarchy of landmarks which have been observed in certain other circumstances.

An even more interesting point is that, as the wasp began to use our landmarks, it showed increasing 'discrimination' between the

187

two types provided—although both had been present for exactly the
same time. Invariably, before a wasp would follow either one or the
other of the two types of landmark, it would follow when we moved
the training constellation as a whole. Also, when two sets of marks
were of almost equal potency, a wasp would, once it began to dis-
regard its own nest, choose both sham nests with more or less equal
frequency. But when the tests were repeated the same day, or during
the course of the next few days, the figures would gradually change:
the wasp would begin to show a clear preference for one of the two
sets of marks which we were using. Of the many instances of this, one
of the most convincing is reproduced below. It concerns wasp 98 in
experiment 31 (upright quarter-cubes $4 \times 2 \times 2$ cm versus half-cubes
$4 \times 4 \times 2$ cm lying flat). The final count was 104:49 in favour of
upright. Training started on July 12th. In consecutive tests the wasp
chose:

14–7 Test No.	Upright	Flat
1	5	8
2	13	13
3	7	6
4	12	2
5	18	4
6	20	8
7	13	1
8	16	7
	104	49

The overall trends were therefore: (1) a gradual change from old to
new landmarks, and (2) progressively more attention to detail—
a continuum from responding to the circle as a whole to responding
to properties of individual landmarks as well.

Finally, we wish to point out, though it is hardly necessary to do
so, that this discrimination is achieved without the aid of either
reward or punishment—on all occasions both the landmarks led
either to a dummy nest, or (when testing was broken off) both to the
real burrow, and there could not have been differential reinforcement.

We therefore had to conclude that this progressive response to
detail in the perceptual field was due simply to repeated exposure to
the stimulus situation.

We mentioned earlier that in tests done at the beginning of training
a wasp would choose both types of landmarks with equal frequency.
The sequence of events was highly characteristic: If there were two

sham nests, I and II, and two types of objects, (a) and (b), placed respectively at I and II, the first choices would be equally distributed between I and II, but soon the wasp would choose, say, I (a) exclusively. If we switched the landmarks, contrasting now I (b) with II (a), the wasp would again choose I. That is to say, from that moment onwards, nest I would be preferred regardless of the marks that surrounded it. In long series of tests it sometimes happened that a sham nest was preferred even when we moved the entire training complex back to the original nest, leaving both sham nests devoid of experimental landmarks.

At first we thought that such a preference for one of the sham nests over even the real nest could be explained by assuming that the situation there resembled our original, pre-training situation (i.e. landmarks not introduced by our training experiment) more closely than that of the other sham nest. But the fact that the preference persisted even after the original training situation had been restored to the real nest, with all its pre-training landmarks, argued against this interpretation. Nor did it tally with the following facts: whenever we noticed a preference for one of two sham nests, we discarded them at once and made a new set of two, in the hope of breaking the 'site preference'. In such a situation a wasp would frequently begin by choosing both new dummy nests with equal frequency, but after a short while a strong preference for one or the other nest would reassert itself. Hence we must conclude that during the course of an experiment, the wasps added *new* landmarks to the configuration which had originally guided them.

This addition of new landmarks to those already in use always seemed to happen quite suddenly. From one moment to the next, the wasp would cease to choose indiscriminately, and at the same time, its whole behaviour changed: hesitation was replaced by fast and firm choices; all at once the animal was completely tied to that particular sham nest.

Furthermore, these sudden preferences tended to occur most often in experiments when we forced the wasps to choose between two types of landmarks of near-equal strengths. The animals found themselves suddenly confronted with *two* nests of equal value in place of one unmistakable nest, and they became disoriented and confused. This sudden utilisation of additional landmarks must therefore be seen as the result of a state of disorientation. The fact that 'wrong' choices occurred even after the training complex had been restored indicates that this is not a case of wasps re-adopting landmarks which they suddenly remembered when they were once more confronted with the familiar arrangement, for if this were so, this late appearance of 'wrong' choices would be incomprehensible. Rather, the animals seemed to switch suddenly to entirely new landmarks of their own choosing, i.e. landmarks chosen without any reward or penalty having been associated with them at all. One could describe these sudden switches by saying that when disoriented a wasp 'speculates' in an arbitrary way.

Krechevsky (9) has described a similar phenomenon in rats. We should like to emphasize the parallel: a rat in a maze will keep changing the cues to which it orients as long as it has not yet learned to use those that guide it to its goal. In the course of a number of days the rats will use different cues one after the other, i.e. they will choose a new kind of stimulus whenever the original one has repeatedly failed to lead them to their goal. This shows that the rat, too, changes the 'landmarks' by which it is guided as long as it is disoriented.

There is another characteristic feature which rats and wasps have in common: the fact that these phenomena are *not* a result of training gradually perfected through repetition, but are, in the words of Lashley (13) sudden 'attempted solutions'. The *sudden* switches in the choice of landmarks described here appear to be quite different from the *gradual* process of learning to use increasingly detailed cues which we discussed earlier. This *sudden* choice of new landmarks occurs only after a phase of non-orientation. Such a phase did not occur when the wasps were slowly getting used to new landmarks.

Finally we have observed indications of yet another sort of change in perception. Under certain circumstances the consistency of a wasp's choice was obviously influenced by extraneous circumstances, making her at one time choose both types with equal frequency, at others show a clear-cut preference.

As an illustration we refer to wasp 103, which was tested for its capacity to discriminate between 'upright' and 'flat' in the manner described in experiment 31.

On August 6, 1937, this wasp, returning with a bee, chose 'upright' 6 and 'flat' 4 times. To obtain even this meagre result we had to prepare new sham nests 3 times in order to eliminate sham nest preference. Some moments later, the wasp dropped its bee, and returned to the nest site unencumbered. It now chose 'upright' 31 times out of 31! This phenomenon was not unique. Wasps were liable to drop their bees when there was a fairly strong wind, or when an experiment went on for a long time, and once the load had been 'jettisoned' their choices suddenly became clear-cut. It was as if a strong effort in the face of repeated frustration impaired the wasp's ability to discriminate.

11. Kühn's Concept of 'Mnemotaxis'

Kühn's attempt (12) to classify the various mechanisms of orientation in animals on the basis of the link between sensory stimulus and subsequent movement led him to a tentative distinction between four types of topotaxis: tropo-, telo-, meno- and mnemotaxis. If we had to put our wasps' orientation to landmarks into Kühn's framework we would of course describe it as mnemotaxis. But the original concept of mnemotaxis suffers from a certain degree of inconsistency inasmuch as it is based on criteria which are quite different from those of the other three types of taxis. This is probably why the concept was not mentioned in Kühn's subsequent (12a) discussion of forms of taxes, and why it was not included in Koehler's (8) survey of orientation movements in animals. Fraenkel (3) goes even further, saying: 'It was found that mnemotaxis is not a special type of taxis comparable to the other types.' However he does not cite any evidence for his statement.

Nevertheless we believe that some of the cases listed by Kühn under the heading of mnemotaxis, and also the achievements of our wasps are a form of orientation which deserves to be considered in its own right.

If we try to analyse the concept of mnemotaxis, we find that the examples cited by Kühn are all characterised by the fact that the responses to the cues used are not innate, but acquired by experience; and also that, in contrast to the other forms of taxis, the movements shown are not linked to sensory stimuli in any rigid way. Information as to the developmental origin of the oriented behaviour, i.e. whether it is innate or acquired, does not of course tell us anything about the ultimately functioning *mechanism* governing it. Kühn himself included, under his examples for mnemotaxis, a case of menotaxis, namely the orientation of a *Lasius niger* which uses the sun (pp. 50–51), though it differs from other menotactic movements only in one respect: in this case we *know* that the ant's capacity for keeping a given direction may remain undiminished for several hours, whereas in other cases, defined as menotaxis, we do not actually *know*, but merely have good reason to assume that this is so. There may be no difference in principle between the 'tendency to persevere' which is present in each case of menotaxis, and 'memory', or 'mneme'.

There is therefore no doubt that memory can be involved in menotactic and, for that matter, any other type of orientation. Yet some other aspects of the examples listed under 'mnemotaxis' deserve closer scrutiny.

As we have seen, the homing of *Philanthus* is not based on movements directed towards, or in a fixed relation to, any single object,

THE ANIMAL IN ITS WORLD

but on movements controlled by entire configurations of landmarks. Moreover, the wasps follow flight paths during which their position relative to these configurations changes from moment to moment; neither the direction from which they are seen nor their distance remains constant; and there can therefore be no question of a constant retinal image, such as, for instance, in the 'optomotor responses' as analysed by Gaffron (4) and Hertz (5). If Fraenkel's claim that 'mnemotaxic orientation is nothing but a sequence of telo- and/or menotactic reactions' were right, one would expect the wasps to follow, on successive visits, more or less constant courses in relation to the landmarks. But this is definitely not the case.

Our first example concerns the flight patterns of a wasp which had learned, without any interference on our part, to use a complex of landmarks consisting of 3 pine cones and 1 twig. We faced this wasp with a number of combinations of some, or all of these objects, leaving the rest of the environment undisturbed (Fig. 56). In this case, as in countless others, the course taken by the homing wasp was never the same, but differed considerably from one visit to the next, even when the markers had not been shifted. When we did move the landmarks individually (i.e. changed their relations to each other) the course of the wasp's flight was entirely independent of the position of any particular one of them.

These irregular flight patterns could of course be seen as a form of searching behaviour, i.e. as unoriented movements, and actually this is probably what parts of them were. But before the wasps alighted, their flights were once more oriented, for their choice of site was strictly governed by the landmarks. When all 4 objects were shifted to the sand patch on the right, our wasp invariably followed. When only 1, or 2, or even 3 were moved, it always went to its own burrow; presumably this was partly due to the fact that the nest was characterised by other marks in addition to those moved by us, but this does not affect our argument. It is interesting to note that the direction from which the animal comes down to land, though somewhat dependent on the direction of the wind, may vary by 90° or more, so that the position of the retinal image can vary even at the moment of the final check.

When there is little wind, this direction may show even greater variation, as can be seen in Fig. 57, which shows the responses of a wasp that had been trained to a circle of pine cones and was in the actual test confronted by two separate half-circles. Here too there is no question of the wasp's taking up a definite position in relation to the landmarks.

To sum up: these examples are representative of what we observed

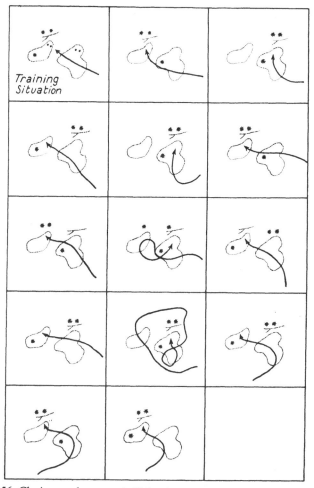

Fig. 56. Choices and approach flights in a variety of landmark arrangements.

throughout: the flight paths of the wasps were in no way fixed; they were variable to a degree which excludes the possibility of their being mere sequences of meno- or telotactic reactions.

Summary

Homing females of *Philanthus triangulum* orient using complex con-

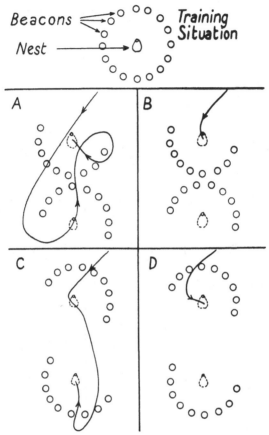

Fig. 57. Choices and approach flights with broken-up pine cone circle.

figurations of visual landmarks, which they learn during their 'locality studies'. Different objects round the burrow are not equal in their suitability as landmarks, and the wasps show distinct selectiveness in what they learn. This study demonstrates their preference for:

Patterned over uniformly coloured flat objects;
Three-dimensional over flat objects;
Three-dimensionally patterned over smooth objects;
Large over small objects;
Objects close to the burrow over equally large distant objects;
Objects further away from the burrow over those nearby if both
subtend the same angle;

194

Objects contrasting with the background over those matching its colour;

Objects present at the first departure over objects added later.

By far the most important characteristic of a three-dimensional landmark is its height above the ground; three-dimensionality as such, or shadows are used to a much lesser extent.

Increasing familiarity with landmarks through repeated exposure leads to increased response to detail; this is a gradual process.

When faced with a disorienting situation a wasp may suddenly switch to responding to landmarks not used before; this behaviour is similar to that observed in disoriented rats.

Kühn's concept of 'mnemotaxis' is discussed, and the differences between orientation to complexes of landmarks and other taxes are emphasised.

REFERENCES

1 BOUVIER, E. L. (1900). 'Le Retour au Nid chez les Hyménoptères Prédateurs du Genre *Bembex*', *C. r. Soc. Biol. Paris*, **52**, 674–6.

2 FABRE, J. H. (1879). *Souvenirs entomologiques*, Ire, 2nde et 4me Série, Paris.

3 FRAENKEL, G. (1931). 'Die Mechanik der Orientierung der Tiere im Raum', *Biol. Rev. Cambridge philos. Soc.*, **6**, 37–87.

4 GAFFRON, M. (1933). 'Untersuchungen über das Bewegungssehen bei Libellen- larven, Fliegen und Fischen', *Z. vergl. Physiol.*, **20**, 299–338.

5 HERTZ, M. a (1929). 'Die Organisation des optischen Feldes bei der Biene I', *Z. vergl. Physiol.*, **8**, 693–748; b (1930). II, *Z. vergl. Physiol.*, **11**, 107–45; c (1931). III, *Z. vergl. Physiol.*, **14**, 629–74.

6 —— (1934). 'Zur Physiologie der gesehenen Bewegung', *Biol. Zbl.*, **54**, 250–64.

7 HÖBER, R., (1931). *Lehrbuch der Physiologie des Menschen*, 6. Aufl., Berlin.

8 KOEHLER, O. (1931). 'Die Orientierung von Pflanze und Tier im Raume, II', *Zool. Teil. Biol. Zbl.*, **51**, 36–58.

9 KRECHEVSKY, I. (1932). ' "Hypotheses" versus "chance" in the pre-solution period in sensory discrimination-learning', *Univ. California Publ. Psychol.*, **6**, 27–44.

10 —— (1932). 'The Genesis of "Hypotheses" in Rats', *Univ. California Publ. Psychol.*, **6**, 45–64.

11 —— (1932). ' "Hypothesis" in Rats', *Psychologic. Rev.*, **39**, 516–32.

12 KÜHN, A. (1919). *Die Orientierung der Tiere im Raum*, Jena.

12a —— (1929). 'Phototropismus und Phototaxis der Tiere', *Handbuch der normalen und pathologischen Physiologie*, Bd. **12**, 1. Hälfte, Rezeptionsorgane II, 1, S. 17–35.

13 LASHLEY, K. S. (1930). *Brain Mechanisms and Intelligence*, 2nd impr., Chicago.

14 OPFINGER, E. (1931). 'Über die Orientierung der Biene an der Futterquelle', *Z. vergl. Physiol.*, **15**, 431–87.

15 TINBERGEN, N. (1932). 'Über die Orientierung des Bienenwolfes (*Philanthus triangulum* Fabr.)', *Z. vergl. Physiol.*, **16**, 305–34.

16 VERLAINE, L. (1924). 'L'Instinct et l'intelligence chez les Hyménoptères. I. Le Problème du Retour au Nid et de la Reconnaissance du Nid (*Vespa vulgaris*

L., *Bombus lapidarius* L. et *B. hortorum* L.)', Mém. (8°), *Cl. Sci. Acad. roy. Belg.*, **2**, 1–71.

17 WEYRAUCH, W. (1936). 'Untersuchungen und Gedanken zur Orientierung von Arthropoden. 7. Teil: Über starke und schwache Orientierungsreize', *Rev. Suisse Zool. Genève*, **43**, 455–65.

18 WOLF, E. a (1926). 'Über das Heimkehrvermögen der Bienen. (Erste Mitteilung)', *Z. vergl. Physiol.*, **3**, 615 bis 692; b (1927). '(Zweite Mitteilung.)', *Z. vergl. Physiol.*, **6**, 221–55.

(From the Zoological Laboratory of Leiden University)

5
The Courtship of the Grayling
Eumenis (= *Satyrus*) *semele* (L.)
(1942)[1]

Introduction and Theoretical Considerations

This work arose from our conviction that in the present state of
comparative behavioural studies we need descriptions and experimental analyses of the behaviour patterns of as many different species
as possible. Numerous studies of the behaviour of birds have been
published and from these many fruitful working hypotheses have
recently been developed, but similar studies of invertebrates are still
rare. Yet such studies might well broaden the scope of these new
working hypotheses. The wide range of behaviour patterns shown by
the Grayling butterfly suggested this species as an especially promising subject.

From the start, we directed our attention towards detailed analysis
of an 'innate releasing mechanism'. An excellent possibility for this
was provided by the most conspicuous response of the Grayling:
when the male is ready to copulate he flies upwards towards a female
passing overhead. It was evident from the beginning of the study that
males fly not only towards females but also towards many other
animals and objects, and from this it was deduced that the 'approach
flight response' requires, and must be controlled by, a very unselective stimulus pattern. The experimental study of this 'releasing
stimulus situation' and especially the question of how its effect compared with a 'simple reflex', seemed worth pursuing.

Later we expanded our investigations to include other behaviour
patterns, and to analyse some of them.

[1] We acknowledge the assistance given by Th. Alberda, D. Caudri, E. Elton,
B. Kok, and P. J. Nieuwdorp. Our sincere thanks are due to Ir. A. E. Jurriaanse
for his permission to conduct our work on his estate near Hulshorst. Drawings
by B. J. D. Meeuse, photographs by R. J. Van der Linde, R. A. Maas Geesteranus
and N. Tinbergen.

Owing to the broad scope of the study and the dependence of our species on the weather (the animals are only active in full sunlight), we were unable to carry out more than a fraction of our plans. Thus we are presenting here only the most reliable of our results. After a description of the general behaviour, especially the male's courtship of the female, the results of two analyses are given: firstly that of the 'approach flight' mechanism and, secondly, an investigation of the

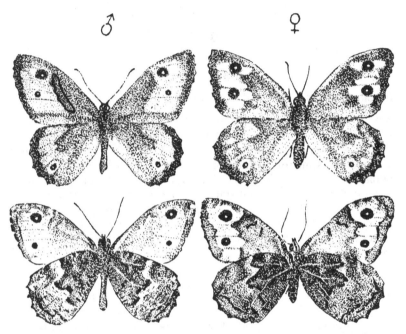

Fig. 58. *Eumenis semele* (L); above, dorsal view; below, ventral view; left, male, right, female. The scent area on the male's left wing is outlined in black.

function of the so-called 'scent patches' on the forewings of the males. Further investigations on the behaviour of this species are in progress.[1]

Coloration

In flight, *Eumenis semele* is a relatively brightly coloured animal. The basic colour while flying is the brown of the upper surface of the

[1] Author's note: As happened so often in studies undertaken jointly with undergraduates, a great deal of information, acquired later, never reached a stage which justified publication.

Fig. 59. Top: a flight cage in a sheltered position. Bottom: rest position of female Grayling.

wings. (Fig. 58). The distal half of each wing contains an ochre-yellow area surrounded by a brown margin. This yellow area is more pronounced in the forewings than in the hindwings, where it is overlaid with a brownish tinge. In the female it is much more brilliantly coloured than in the male. The 'eye spots' in the female are also much blacker and their centres more striking than those of the male. The male possesses a patch of dark scales of a different colour on the upper side of the forewing. We shall deal later with the function of these patches.

The coloration of the underside consists of two main areas. The ventral coloration of the forewings is similar to that of the upper surface. The hindwings, by contrast, are clearly procryptic; they have a 'marbled' pattern of fine and coarse grey-brown stripes composed of separate areas of white, golden-yellow and dark brown scales. These form very irregular oblique bands. Across the middle of the wing, this delicate pattern is overlaid by a distinctive wavy light band, bordered by darker colour on its proximal side. This dark border stands out clearly against the white and gradually fades out towards the outer wing edge. The white oblique band is much more striking in the male than in the female.

This intricate pattern of fine and coarse banding, together with the general grey coloration, renders the animal almost invisible to human eyes on backgrounds such as grey sand, brownish moss, dead leaves or bark. The behaviour of the animal also fits in well with this camouflage. When resting, the wings are held close together so that only the cryptically coloured parts are visible. On the forewings, the cryptic coloration is confined to the parts visible in the resting animal (Fig. 59). Furthermore, the animals either move rapidly or not at all (except while searching for food or courting). Hence one usually loses sight of them as soon as they settle. Many cryptically coloured grasshoppers behave in a similar manner; these 'dash-and-freeze' tactics are, certainly, a behavioural component of camouflage. That this sudden immobility after rapid movement has a definite effect on predators that hunt by sight was shown by Lorenz (15) who found that a jackdaw's attention was always drawn to leaping grasshoppers but that it was unable to relocate them after the jump. However, we have no observations on the effectiveness of the cryptic coloration or behaviour of *Eumenis*.

Description of Behaviour Patterns

Occurrence In the Netherlands *Eumenis semele* inhabits dry sandy areas of either alluvial or diluvial origin. In our experimental area, the diluvial sand zone of the Veluwe in Gelderland, this species is found in association with certain grasses typical of such poor soils, mainly *Corynephorus* (= *Weingaertneria*) *canescens* (L.) and *Agrostis canina* (L), which are the food plants of the caterpillar. The freshly hatched imagines are found only in places where these grasses are growing, between bushes or trees which protect them from the wind. In completely open areas of the inland dunes, which have a gravel substrate produced by wind erosion, both grass species occur but the habitat is more exposed and in such areas no freshly hatched animals can be found. Obviously the egg-laying females select not only the suitable food-plants but also a sheltered habitat. If, on different days, the wind comes from different directions, the females show a suitably adjusted distribution; they are always found in places that are overall most sheltered. The avoidance of open spaces is of survival value, in that it reduces the chance that the wings of the emerging imagines will start to dry before being fully expanded. On two occasions, both in very strong wind conditions, we have actually found misshapen animals whose wings had dried in a half-expanded state. The probability of such accidents is, of course, greater in windy places.

As has been reported occasionally, the males of this species emerge earlier than the females. In the first few days of July it was indeed always very striking that females were still rare when males were already fairly plentiful. On most sunny days in 1939, 1940 and 1941, we counted the number of males and females over roughly the same circuit through the experimental area (Fig. 60). Since the total number seen varies with the weather and also with the duration and course of our excursions, the absolute numbers do not mean very much. However, the gradual increase in the proportion of females to males shows up very clearly.

The places where the Grayling hatches generally provide no food for the adults, who live on nectar. Accordingly, once the wings are dry (usually on the day of hatching) the animals leave in search of food and in doing so, may travel quite considerable distances. One can often see such wandering individuals flying over the bare inland dunes. The distance covered in these 'exodus flights' is certainly often greater, perhaps on occasion even much greater than 1,200 m, but of course the chance of recording their extent decreases with distance.

To obtain more accurate information we colour-marked (particularly in 1938) a fair number of individuals with poster paint mixed with alcohol-based shellac or, we painted a clearly visible number under each hindwing, using aluminium paint. These marking attempts yielded only sparse data. The longest time between release

201

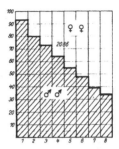

Fig. 60. Percentage frequency of the sexes in successive weeks during
the breeding season (2,086 sightings over three years).

and resighting was 42 days; the greatest observed displacement, as stated, was 1,200 m. Approximately 80 per cent of the marked animals were not seen again. Since the movements appeared to be completely non-directional and return of even the females to their area of origin could not be proved, one cannot (as yet?) call this 'migration'. Nor was there any sign of attachment to one place, though some animals might be found for a number of days on the same patch of flowers. Even this was observed in only a small proportion of the animals. When one continues to observe and mark over a period of days at the same food source, one soon sees that the population is in a state of flux.

Feeding In the course of July the numbers of animals seen at the feeding sites increases. Two very different types of food source are visited, distinguished by the butterflies' method of finding them. '*Charaxes* type' of response: In the very dry areas the animals congregate on 'bleeding' birches or oaks, that is, trees which have been attacked by *Cossus cossus* (L) larvae, and which 'bleed' a strongly smelling sap. Many insects drink from such sap-oozing trees: *Cetonia aurata* (L), various Vanessids (*Pyrameis atalanta* (L), *Vanessa polychloros* (L), *V. antiopa* (L), *V. io* (L), *Pararge aegeria* (L), var. *egerides* (STGR), *Agrotis* spp., many flies, honeybees, *Vespa* spp., ants etc. *Vespa crabro* (L), *V. media* (De Geer) and *Mellinus arvensis* (L) look for the sap trees mainly in order to catch flies, and the two *Vespa* species also catch butterflies (*V. antiopa*, *V. io*, *Pararge aegeria* and also *Eumenis semele*).

Eumenis semele detects the trees by smell. One can often observe flying animals become suddenly alerted when they pass a sap tree on the lee side; when flying past, they are often caught in the scent clouds wafted away from the tree by the wind. Here they change their behaviour abruptly, they brake, and fly slowly upwind, oscillating in

202

a horizontal plane, until they reach the source of the scent. We did not analyse this response further. The similarity to the behaviour of *Charaxes jasius* (L) Knoll (**14**) is striking.

The animals also learn to orient visually towards a tree where they have previously fed. One can detect such 'regular customers' since they fly directly towards the tree from any direction. Dependence on the direction of the wind is either not observable in these individuals or, if present, is not very pronounced.

The approaching animals settle either on the sap birch or in its immediate vicinity, and while completing the last part of the approach, show continuous movement of their antennae. The antennae almost certainly possess chemoreceptors; we also saw indications that, as in *Pyrameis atalanta* (Minnich, **17**), the tarsi on the second pair of legs might bear contact chemoreceptors. We therefore offered to different individuals of *Eumenis semele*, *Epinephele jurtina* (L) and, as a control in case of negative results, *Pyrameis atalanta*, after previous satiation with water, the following substances in concentrated aqueous solution: arabinose, fructose, galactose, glucose, isoduclite, lactose, mannitose, raffinose, saccharose, xylose and common salt. The animals were held by the wings in a wooden clamp and the tarsi were brought into contact with the water, following which the tongue was extended. Once they were fully satiated, they were left undisturbed for two minutes. Water was then offered once again and only when this attempt proved negative were the test substances presented.

Whereas *Pyrameis atalanta* often gave positive responses i.e. tongue extensions to sugary solutions we were never able to elicit any response in the other two species once they had ceased to react to water. Thus, all species are able to sense water with the tarsi but only *P. atalanta* possesses a tarsal chemical sense.

Apart from the *Cossus*-infested birches, *E. semele* also seeks out other strongly smelling food sources: horse droppings, lime (*Tilia*) flowers, sweat and urine from horses and men, etc.

Vanessa urticae type of response *Eumenis semele* searches just as commonly for flowers, which it doubtless detects visually. We observed them sucking nectar from *Erica tetralix* (L), *Calluna vulgaris* (Salisb.), *Epilobium angustifolium* (L), *Cirsium* spp., *Jasione montana* (L), *Lotus uliginosus* (L) and yellow Compositae. Frohawk (**9**) also lists, apart from *Rubus* and *Cirsium*, *Erytraea centaureum* (Pers.); Eckstein (**7**) mentions *Thymus*; Heimans (**10**), *Genista tinctorium* (L).

When observing the Grayling flying from flower to flower, one immediately suspects that these flights are visually directed. The fact that if one covers individual flowers with glass cylinders, the animal will fly against the glass supports this suspicion. Since, in connection with later experiments on the stimuli eliciting the sexual approach flight response of the males, it was important to determine whether *E. semele* can distinguish colours, we gave special attention to their orientation to flowers. We proceeded in two ways:

(1) We kept a number of animals in a cage measuring $5 \times 2 \times 2$ m (Fig. 59; top), on the floor of which we laid out differently coloured, as well as grey, Hering papers. The Graylings did not fly towards these papers unless there were some scent present, which is in accordance with what Ilse reported of other butterflies (**12**). For this purpose, we used a variety of flower oils (carnation, peppermint,

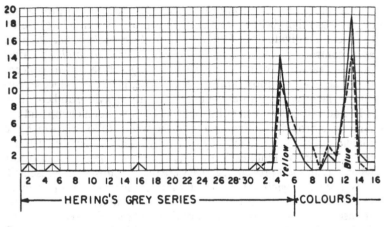

Fig. 61. Number of approach flights by *Eumenis semele* to the Hering grey and to coloured papers. Ordinate: number of approach flights observed; abscissa: the papers used.

rosemary, almond, rose, jasmine, etc.) which we did not, as Ilse did, spray into the air, but put on cloths which were hung on the wind-ward side of the cage. The animals reacted immediately, but by flying towards the coloured papers. Thus the scent clearly had an activating, but not a directing effect. It was found that the paper sheets had to be rather large; the best results were obtained with sheets of 12×20 cm placed on a white background. In these experi-ments, the results which are given in Fig. 61 (dotted line), we utilised only a few shades of grey.

(2) In one place in the field where the Grayling were searching for *Epilobium angustifolium* and yellow Compositae (*Hieracium* spp.), we put artificial flowers of Hering paper among the real flowers.

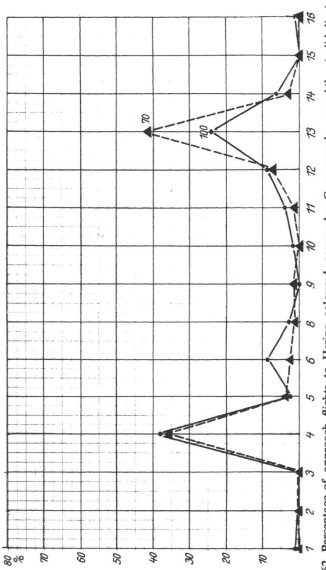

Fig. 62. Percentage of approach flights to Hering coloured papers by *Coenonympha pamphilus* (solid line) and *Epinephele tithonus* (dotted line). Ordinate: approach flights observed; abscissa: the papers used.

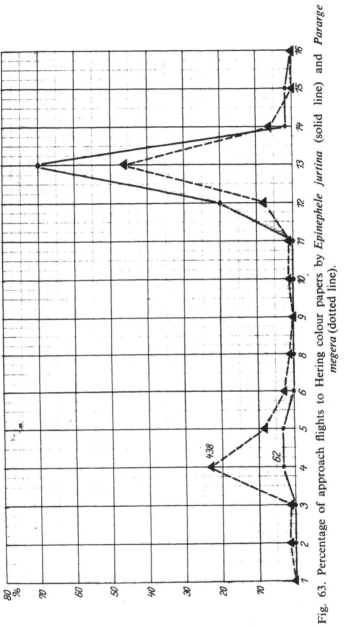

Fig. 63. Percentage of approach flights to Hering colour papers by *Epinephele jurtina* (solid line) and *Pararge megera* (dotted line).

Here we used the 16 colours of the Hering colour series as well as all the 30 greys. The results of these experiments are shown by the solid line in Fig. 61.

It is clear that while searching for flowers, *E. semele* orients mainly towards yellow and blue. The same experiments were then done with the related species *Epinephele jurtina*, *E. tithonus* (L), *Pararge megera* (L) and *Coenonympha pamphilus* (L). All these species are attracted to blue and yellow, so showing the 'Vanessa type' of response described by Ilse (Figs. 62, 63 and 64). *Pararge*, however, responds

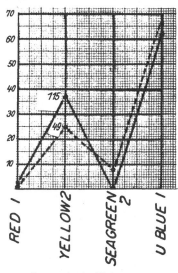

Fig. 64. Percentage of approach flights to four Oswald papers by *Epinephele jurtina* (solid line) and *Eumenis semele* (dotted line).

mainly towards the blue paper. The preference for yellow and blue concurs well with the observation that *E. semele* is found most often on yellow, blue and blue-red flowers.

Defence Reaction *Eumenis semele* shows two kinds of response towards enemies. Small animals are warded off by a rapid flapping of the wings. At higher intensities, this behaviour is rhythmic, the wings being flapped open and shut 5 or more times in rapid succession. In this way, *E. semele* drives away flies and wasps, for example, from the *Cossus* birches; and a female which has already been fertilised defends herself in this way against males attempting to mate (Fig. 65). This intense, repeated defence behaviour was observed most often in response to such obtrusive males.

Fig. 65. The defence response of a female (right) unwilling to copulate with a courting male.

Larger animals elicit flight. In separate experiments, to be described later, we shall show that the smallest object eliciting this response measures approximately 8 cm in diameter at a distance of 50 cm.

The response has two stages. The first, a 'startle' response, revealing the intention to flee, is a sudden lifting of the forewings, by which the black, white-centred 'eyespot' is exposed. Once an animal has taken up this position (Fig. 66) it is extremely sensitive to all flight-releasing stimuli. Whether this 'intention movement' also functions to ward off attackers is not certain, although it seems likely, since the animal is showing conspicuous structures (the 'eyespots') which contrast vividly with the cryptic coloration, and the movement is strictly limited to the display of this structure.

The second stage is the actual flight. In the beginning, this is oriented only in so far as it is leading away from danger, without however being aimed towards any hiding places, or towards light. But after continued disturbances the animal has a tendency to make for the tops of trees. Very severe disturbances, e.g. catching or rough

Fig. 66. Exposing the wing 'eyespots'.

handling, result in the animal's flying towards the light immediately on release. Animals blinded on one side fly in circles towards the unblinded side, as we accidentally observed in animals blinded for other purposes. The flight towards light is thus tropotactically oriented.

Basking Especially in cool weather (and on warm days preponderantly in the morning and evenings), the butterflies single out sunny spots where they settle with the longitudinal axis of the body approximately at right angles to the incident light. If the sun is not too low, this orientation movement is accompanied by a tilting of the whole body, which exposes one side approximately at right angles to the sun's radiation. Thus the side exposed to the sun receives the maximum radiation. The head position, however, remains normal, i.e. symmetrical with respect to the vertical.

By this method of sunning the protective coloration retains its full effectiveness. A spreading of the wings, such as can be regularly observed, for example, in the *Vanessa*s and also in various relatives of *Eumenis* such as *Epinephele tithonus* and *E. jurtina*, would have, without doubt, a greater warming effect but would at the same time render the protective coloration of the underside ineffective. *Eumenis semele* never spreads the wings while sunning.

The fact that this behaviour occurs only in cool weather with sunshine indicates that temperature plays a part in its release. But in its orientation, light may well be important: with thick cloud cover it may happen that a relatively light patch of sky appears quite a distance from the sun, and then the butterflies turn their broad side towards this patch. If one illuminates a completely shaded animal with sunlight reflected from a mirror, it often orients itself to the reflected rays. The possibility of other stimuli, e.g. temperature stimuli, having an orienting effect has not been investigated.

If the animal is shaded, it will move out into the sunlight. In cool weather this reaction is immediate, in warmer weather it is shown after 1 sec to 5 min., sometimes even longer. The fastest response occurs when the animal is completely shaded; shading of the head alone is also effective; but even shading of the thorax and wings, with the head still in the light, is likewise followed by resettling, though less promptly—varying from a few seconds to even minutes. Experiments to further determine the roles of temperature and light stimuli were started but not completed.

Courtship Males perform sexual movements even before they have seen females. As early as July, they respond to passing butterflies of

209

any kind by an immediate and surprisingly fast pursuit. But such 'erroneous' flight responses are not maintained; as soon as a responding male has reached the other butterfly he will abandon it. At the sap trees there are often many, up to ten or more, of these sexually motivated males flying not only at butterflies of other species, but also at each other, and this often results in the whole group being stirred up.

The function of this flight is obvious once the females appear, for then it is followed by a complicated courtship. The full sequence is not easily observed in the field; to supplement our few, and often incomplete field observations, we therefore arranged a series of observations on animals which we kept in large flight cages.

When a male flies towards a receptive female, she will sooner or later, sometimes immediately, settle on the ground. The male settles near, and usually behind her, and then proceeds by short jerky movements around her side until he is facing her. This 'circling' involves a series of rapid sideways steps interspersed with pauses, and takes about 3 sec. Then the actual courtship begins with 'wing quivering'. The male's forewings are slowly raised several times with slightly spread frontal margins, then quickly closed and lowered. Usually two such movements occur per second. Each movement is a little more pronounced than the previous one and the forewing margins are spread a little further each time. By the time of the last occurrence, approximately half the forewings protrude from the hindwings. A bout of wing quivering is usually followed by two other movements, which occur simultaneously: 'fanning' and 'antenna spinning'.

In fanning the forewings are kept in a raised position. Their leading edges are separated from each other and closed again rhythmically and at a fast tempo. Only the frontal one-third of the wings are opened and they separate no more than approximately 1 mm.

Antenna Spinning During quivering, the antennae of the males are spread horizontally and at right angles to the longitudinal axis of the body. As soon as quivering develops into fanning, the antennae begin to perform a remarkable movement. Held completely stiff, they make a conical sweep so that the tips describe a circle. Both antennae move synchronously and in the same direction, that is, back–upwards–forwards–down. Gradually, this circle becomes an ellipse with its longitudinal axis inclined forwards and upwards, making an angle of 45° with the ground. The downward movement is clearly faster than the upward one. Also, the antennae are gradually directed forwards. We observed one male in which the antennae

Fig. 67. Bowing. Male on the right.

rubbed against the legs of the female, but this is by no means the rule. Each complete circling lasts about ¾ sec. A bout of wing fanning and antenna spinning can last from 1 sec to several minutes. Fanning is the first to cease when courtship lapses, e.g. when a cloud covers the sun; its tempo becomes slower and finally antenna spinning is also completely suspended.

The last component of the actual courtship is 'bowing' (Figs 67 and 68). In this spectacular display the male spreads both pairs of wings and brings the forewings very far forward so that they are completely separated from the (likewise somewhat raised) hindwings. Once in this forward inclined position, the male closes the wings again extremely slowly—as it were, 'with emphasis'. In most cases, this catches the antennae of the female between the male's forewings. As soon as the male has shut the wings, they are drawn backwards. The whole movement lasts about a second and is a perfect and elegant finale.

After bowing the male attempts to copulate. He accomplishes this by moving quickly sideways round the female (Fig. 69). While doing this, he already bends the abdomen forwards and to the side so that the copulatory apparatus, which protrudes from the end of the abdomen like a brush, is pointed forwards to enable it to be hooked into the female's genitalia. We shall not describe here how the individual claspers etc. move. Curiously, it occasionally happens that a male will bend its abdomen to the wrong side, i.e. to the right when the female is on its left side, or the other way round. As soon as the

♂   ♀

Fig. 68. Bowing seen from above.

male has hooked fast, he turns the front of his body until the whole body is again extended and he is in the same plane as the female. The two animals then remain in the well-known butterfly copulatory position.

The attempt to clasp often follows immediately after the first bow. The sequence of components is usually as given above: quivering, fanning and antenna spinning, bowing and clasping. Sometimes the courtship display is repeated one or more times and the sequence then becomes somewhat irregular: thus quivering in particular, or indeed any of the behaviours preceding clasping can drop out.

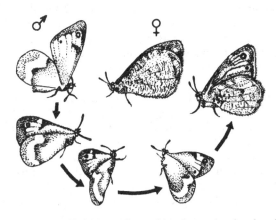

Fig. 69. The half-circle performed by the male after bowing.

As an example of a long and rather irregular courtship sequence (which one can expect in a flight cage, or in the field with an unwilling female or in cool weather) we present the following observation from July 18, 1940.

Male No. 3 courts female H in the flight cage. It performs as follows: Antenna spinning (AS) and Fanning (F), Quivering (Q), AS-F, Bowing (B), Copulation Attempt (CA), back to Facing Opposite (FO), AS-F, B, AS-F, B, CA, FO, Q, AS-F, Q, AS-F, B, Q, AS-F, B, Q, AS-F, B, Q, CA, FO, Q, AS-F, B, CA, FO, AS-F, Q, AS-F, Q, B, Q, B, AS-F. Then a cloud covers the sun, and a long period of AS follows; then the male finally becomes completely motionless. In the wild, such a long series is the exception; copulation often succeeds at the first attempt. Our captive females were often cornered, whereas a free-flying female can always fly off.

The behaviour of a receptive female during courtship is very simple. She usually stays completely motionless in one place. Only when the male begins his half-circling after bowing does the female raise the wings a little so that the posterior genitalia are freed. If the female does not behave quietly, she is obviously less stimulating to the male; sometimes the courtship may even be completely broken off. This is particularly noticeable when the female responds with the defensive wing flapping mentioned earlier. We observed this inhibiting effect even with males that were so strongly motivated sexually that they attempted to copulate with other males. Occasionally, males continue stubbornly to court even wing flapping females.

Copulation lasts from 45 min to 2 hr. At the beginning the genitalia of the male make regular 'pumping' movements every second, later the rhythm becomes slower. We did not investigate what is actually

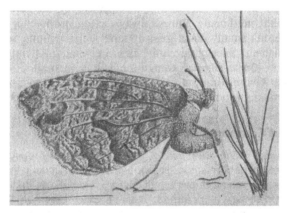

Fig. 70. Ovipositing female (from a photograph).

occurring here—the movements might either serve to achieve a complete hooking-in or to transfer sperm.

When a copulating pair is disturbed, the female flies off, the male letting itself be carried passively with folded wings. A fertilised female is most probably not prepared to mate again. She responds to each approach of a male with wing flapping. By contrast, the males seem to be able to pair several times. We have no evidence for this from field observations but on several occasions in our flight cages we saw a male courting a female again one hour after a successful copulation.

Oviposition This follows shortly after copulation; how soon we cannot state exactly. Towards the end of July, one can already see many females in the areas suitable for oviposition. These are, as said before, somewhat protected places where grasses, especially *Cory-*

nephorus canescens and *Agrostis canina* are growing. The females search for these areas especially between 11 and 14 hrs (sun time). How they locate them we do not know. We have often observed oviposition itself. During this, the female proceeds with a high-stepping walk through the grass clumps, stops suddenly, curls the abdomen (Fig. 70) and touches a section of the plant directly in front of her with it. The abdomen is then usually drawn back, pressed against a new section of plant etc. until finally the egg is placed in this way on one of the tested sites. The animal then moves on and soon lays another egg. Sometimes a female will fly for a short distance between two egg depositions.

It appears, therefore, that the females test the substrate in some way by probing it with the point of the abdomen. We have not investigated whether this testing is chemical or tactile. But it is striking that the females almost always select, in the end, pieces of *dead* material, usually dead grass or sand sedge sections, sometimes even dry lichens; the eggs are only very rarely laid on living shoots of *Agrostis* or *Corynephorus*, even though it is such shoots that are tested first. Oviposition on dead material could well be a protection against grazing rabbits, mice or voles, or even grasshoppers.

Analysis of Courtship Behaviour

The Approach Flight Response of the Male   As mentioned above, a male moves in an unmistakable manner towards a female flying overhead. The movement of the wings is more hurried, the flight is faster and the trajectory is straighter than in other types of directional flight, as for instance in the approach to a flower. Thus one can recognise the sexual flight by its movements alone. These circumstances made it possible for us to establish that the approach flight is a response that cannot be released in the absence of external stimuli. No doubt, as we shall see, the stimulus threshold can rise and fall under the influence of internal factors but it never reaches the point of a real 'vacuum activity'.

Even simple observation shows that the approach flight is initiated by visual stimuli; the experiments that form the greater part of this work give abundant confirmation of this.

The approach flight is, furthermore, telotactically oriented. We once saw this particularly impressively when observing a number of unilaterally blinded animals. As previously stated, these animals escaped by flying in circles towards the blinded side. However, as soon as a male treated in this way spots another butterfly it will follow it in a completely normal fashion, in a dead straight line.

Scaring such a male by vigorous movements produces the circling movement once again—a textbook demonstration of different movements having different taxis-components.

From direct observations we can state further that an *Eumenis* male flies not only towards females of its own species, but also towards other flying animals. We have seen approach flights directed towards the butterfly species *Eumenis semele, Hipparchia statilinus* (Hufn.), *Epinephele jurtina, E. tithonus, Pararge megera, P. aegeria* var. *egerides, Coenonympha pamphilus, Lycaenea icarus* (Rott.) and other *Lycaena* species, *Cyaniris argiolus* (L), *Chrysophanus phlaeas* (L), *C. dorilis* (Hufn.), *Augiades comma* (L), *A. sylvanus* (Esp), *Plusia gamma* (L), *Lasiocampa quercus* (L), *Pyrameis atalanta* (L), *P. cardui* (L), *Vanessa urticae, V. io, V. polychloros, V. antiopa, Pieris brassicae* (L), *P. rapae* (L), *Aporia crataegi* (L), *Gonepteryx rhamni* (L); and also towards the beetles *Geotrupes vernalis* (L), *Spondylis buprestoides* (L), *Anomala aenea* (De Geer), *Serica brunnea* (L), *Cicindela sylvatica* (L) and *C. hybrida* (L); towards the Hymenoptera *Ammonophila sabulosa* (L), *A. campestris* (Jur.), various *Bombus* species, *Mellinus arvensis* (L), *Philanthus triangulum* (Fabr.), *Vespa rufa* (L), and *Vespa crabro* (L); towards the Diptera *Eristalomyia tenax* (L), *Lasiopticus pyrastri* (L), *Asilus crabroniformis* (L), *Calliphora* spp. etc.; towards the Odonata *Aeschna* spp., *Libellula* spp., *Sympetrum vulgatum* (L), *Calopteryx virgo* (L); towards the grasshoppers *Stenobothrus* spp. and *Oedipoda coerulescens* (L); and even towards birds: *Emberiza citrinella* (L), *Fringilla coelebs* (L), *Parus major* (L) and *Turdus viscivorus* (L).

Apart from these, males often respond to falling yellow, green or brown birch leaves, to fluttering leaves on trees, to pine cones lobbed overhead or even to patches of shade that such objects throw on the ground. Sometimes a male will even fly towards its own shadow. From these observations alone, one can infer that the approach flight response must be dependent on a very simple stimulus pattern. Colour, shape and size can obviously vary greatly. Amongst the animals named, the butterflies are the most definitely preferred and we presumed therefore, even at the beginning, that the characteristic fluttering movement of butterflies must be an important factor.

We realised that here was an excellent opportunity to study, in detail, a 'releasing stimulus situation'. Through systematic recording of the responses of males to a variety of dummies differing in one characteristic after another, and comparing these responses to those elicited by an optimal dummy we could hope to determine accurately the relative stimulating value of each of these characteristics.

The method applied was very simple. Models or 'dummies' were cut from solid paper. Each model was attached by a thread 1 m long to a thin twig of the same length and with this 'fishing rod' we could make our model 'flutter' in a fairly constant and natural way, starting about 3 m away from a male. We then determined whether or not an approach flight response was elicited. It would doubtless have been better if we could have measured the response intensity for each approach flight in some way. This was, however, not possible in this case since the response, when it was elicited at all, was always complete. The explanation for this appeared to us to be the following: we could often show that an animal in flight could be much more easily made to do an approach flight than a settled one. This means that the threshold of the approach flight is higher in a stationary than in an already flying male, and that, once airborne, a male was bound to respond with high intensity. The ultimate outcome was in effect an all-or-none response. We were thus forced to take the ratio between the number of positive responses and the total number of experiments as a measure of the releasing value of the model.[1] This made a large series of experiments necessary. The proportion, expressed as a percentage, we termed the 'approach flight index'.

We confined ourselves exclusively to successive presentations of models and did no choice tests. As we will show later, distance from the model is an important factor and one which was difficult to keep constant. Each model was presented 3 times in succession, usually at 5–10 sec. intervals. Then followed, usually after 15–25 sec, a series of 3 experiments with another model, etc. The sequence in which we presented different models was arranged in random order. The experimenters, on occasion, also changed their 'fishing rods', or only the models, to prevent possible differences between people operating the 'fishing rods', or between the rods themselves, from having an effect. We worked with the same male until it went off in search of food or flew so far away that we could not accurately identify it again. Many series were broken off by a mutual approach response between the experimental male and another male, after which certain identification was often impossible. The total number of experiments performed with a single male was thus determined more by the experimental animal than by ourselves and shows, therefore, wide fluctuations, in fact from 6 to several 100s; this accounts for the untidiness of many of our graphs.

The Optimal Model As an optimal model, which appeared to our eyes to be about the same colour as an *Eumenis* male, we used pieces

[1] It was later discovered (see below) that the approach flight distance could also be taken as a measure of the intensity.

of grey-brown packing paper cut into the shape of an *Eumenis* without any other particular markings (Fig. 71 B). All models which had to have the same colour as this standard ('normal') model were cut from the same piece of paper. This colour was not completely optimal (since, as we shall see, there is a 'supernormal' colour) but it was not inferior to the natural *Eumenis* markings. To demonstrate this, we compared, in two series, this model with a model on to which we had glued natural *Eumenis* wings ('glued' Fig. 71 A) and with a model painted true-to-nature with watercolours ('visual' Fig. 71 C).

The results are given in Fig. 72: all models elicited approach flights with equal frequency. Therefore the typical markings of *Eumenis semele* have not the slightest effect on the release of the sexual approach flight.

The reaction of the males was, at first sight, completely similar towards all three models, but once they had reached them they followed the glued *Eumenis* wing model more persistently. It appears, therefore, that a second response follows the approach flight, and that this 'following flight' is guided by additional stimuli. Since the 'glued' model differed from the standard model not only in colour and pattern but also (chemically) by the presence of the glue, we conducted a new experiment in which the 'glued' model was presented alternately with a standard model smeared with the same glue. Even then the males followed the 'glued model' much more often than the standard model covered with glue. The 'visual model' released no 'following' flight at all. Probably therefore chemical stimuli emanating from the wings of the female (and to a slight degree some component of the glue) are responsible for the following flight, but we did not pursue this question further.

Colour   Once we knew that details of patterning had no effect, we used uniformly coloured butterfly models cut out of Hering paper in the shape of the standard model shown in Fig. 71. The following colours of the Hering series were used: white, black, yellow 4, green 7, red 1 and blue 13. Since it emerged that white elicited fewer approach flights than the other colours, we conducted a parallel series with white, grey 7, grey 15 and black.

Fig. 73 shows the data obtained from a number of experiments. It was seldom possible for us to present the whole series of models; each experimenter could at best handle 2 colours and, since 1 person must always act as recorder, 2 persons usually could work with no more than 4 models and we often used only 3. Hence the heterogeneous data.

The total number of approach flights varies in the different experi-

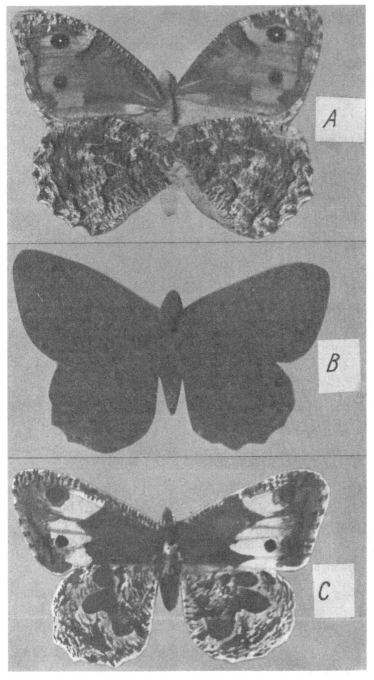

Fig. 71. Three butterfly models as used to analyse the 'following' of the male Grayling. Top: wings of female glued on. Centre: standard (normal) model covered with glue. Bottom: painted model, without glue.

ments, firstly with temperature (lower on cold days, as well as in the mornings and evenings) and secondly with the time of year (the males respond better in August than in July).

As the diagram shows, all the colours used have about the same releasing value. Only white usually lags far behind, especially in experiments where the overall response level is low.

To ascertain whether the differences between models were due, as

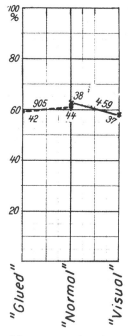

Fig. 72. Results obtained from two series of experiments with the models shown in Fig. 71. Ordinate: percentage of positive responses (see text). The two-digit numbers give the average approach flight *distance* in cm. The three-digit numbers give the number of observations in each series.

we suspected, solely to their brightness,[1] we compared the black and white models with models made of two intermediate shades of grey paper (Hering 15 and 17). The results are given in Fig. 74. Thus the lighter the model, the lower its releasing value. The male could be said to respond to a silhouette.

[1] That the U.V. difference cannot be the decisive factor here, can be determined from Lotmar's (16) publication on the reflection curves produced by Hering papers.

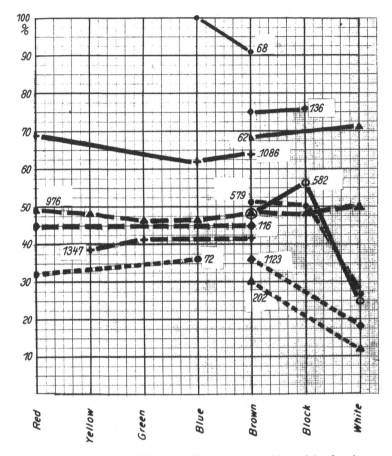

Fig. 73. Twelve series of 'fishing rod' experiments with models of various colours. Ordinate: percentage of positive responses. Numbers on the graph refer to the total number of approach flights in different experiments.

In Fig. 75, the curves of Fig. 61 (responses of feeding butterflies) are shown together with the approach flight curve. Since the uniform brown standard model was used in most experiments it is used again as the basis for comparison and given the index 100. For all other colours only those series in which the standard model had been included were taken into account. The curves show, therefore, the approach-releasing value of the various colours in comparison with that of our standard brown. In all, 6,211 positive responses are

220

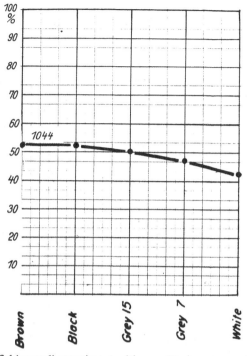

Fig. 74. A 'fishing rod' experiment with grey Hering papers. Ordinate: percentage of positive responses.

represented; a series of 72 using red and blue was discarded since brown had not been included.

The procedure had been too haphazard to warrant statistical analysis of the data, but in view of the massive and obviously quite consistent evidence on the sexual approach flight we feel entitled to conclude that, in contrast to the feeding behaviour, the sexual approach flight is not controlled by any specific colour, but rather by the darkness of the model. This conforms with the results of the grey series.

The differences between the colours are too small to ascertain exactly the releasing value of each.

Fig. 75 shows, further, that the natural colour of the female is not optimal for the approach flight of the male, black having a stronger effect. Black females would therefore be a much stronger stimulus than brown ones. One could thus term the black model 'supernormal' (see further p. 226).

These facts by no means signify that *Eumenis* is colourblind, that is, that the eyes are unable to respond differentially to various colours since, as we have seen, an animal searching for flowers responds very differently to the different colours. That is why we have entered in Fig. 75 the releasing value of colours during feeding,

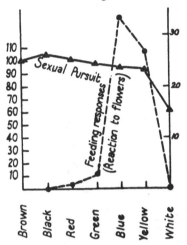

in which the values for the two curves in Fig. 61 are combined; the actual numbers of responses are given at the right of the figure.

We have, therefore, demonstrated what one could call 'central colour blindness' typical of males approaching females, and which contrasts strikingly to the specific sensitivity to particular colours in hungry animals. In other words, different responses of one and the same animal may have different 'releasing mechanisms'. Each response appears to be elicited by a selection of only a very small part of the total potential input from the sense organs. Two inferences can be drawn from this: firstly, that one cannot, by the study of a single response, determine the faculties of the sense organs; and secondly, conversely, one

Fig. 75. Averaged data for the colour experiments; the sexual approach flight (solid line) is compared with searching for flowers (dotted line). Ordinate: left, the sexual releasing value of the colours relative to brown = 100; right, the number of feeding responses. Original data are given in Figs 61 and 73.

can not, on the basis of hitherto acquired knowledge of sensory potentialities, predict which stimuli will elicit any particular response. The first inference was already made by Von Buddenbrock (3) in his treatise on colour blindness; for instance, Schlieper's (19) data indicate that *Hippolyte varians* (Leach), an animal doubtless able to discriminate colour efficiently (as shown by its adaptation to the colour of the substrate) shows its optomotor response only to variations in brightness irrespective of colour; thus, when performing an optomotor response, *Hippolyte* is likewise 'centrally colourblind'.

Shape We presented the shapes shown in Fig. 76 in various combinations, using the standard brown paper. The models had approximately the same surface area; the butterfly model is about natural size.

One problem in assessing the results (Fig. 77) is that altering the shape of a model also altered its movement—when doing these experiments we were not yet aware of the effects of the *type* of movement. We were later able to demonstrate that a fluttering or dancing model was a much stronger stimulus than one moving smoothly. In our 'fishing rod' experiments the model both 'danced' and revolved round its axis. The dancing movement was least pronounced in the longest rectangle, but it revolved more rapidly than the other models, probably too quickly (even though butterflies undoubtedly have a very high 'flicker frequency).' That is why, to us, the longest is much less butterfly-like in its movement than the other models. The low score of this model may therefore have depended, at least in part, on its movement. It is, however, improbable that it was entirely due to deficiencies in movement (see Fig. 83).

Fig. 76. Models used in the shape discrimination experiments.

It appears to us that these objections are not strong enough to invalidate the data, for we were less interested in shape discrimination than in the apparent lack of it; at any rate in the lack of selectiveness in this particular response. This is why we decided not to test differently shaped models which could be moved in an identical manner.

Size The effect of size was studied in various ways since it presented more complicated problems than did colour or shape.

We began by presenting the series of butterfly-shaped models of different sizes shown on the right of Fig. 78, the largest being $4\frac{1}{2}$ times and the smallest $\frac{1}{6}$ the normal size. The results are given in Fig. 79. Natural size was obviously optimal, more than a few approach flights were, however, made even towards models at the

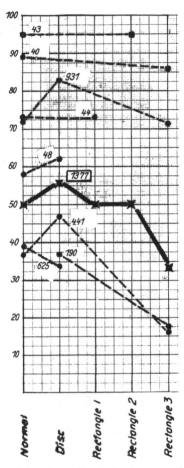

Fig. 77. Results of the shape discrimination experiments. Thick solid line = average. Ordinate: percentage of positive responses.

extremes of the series. As we shall discuss below, the scores obtained in these experiments were complicated, especially with the largest models, by the involvement of escape tendencies. The curve of Fig. 79 falls off sharply for the larger models; as we shall show later, it should, if escape tendencies had not been involved, rise rather than fall.

Fig. 79 is also unreliable firstly because it is based on comparatively few observations, and secondly because it was derived from series in which at the most two, or rarely three or more, models were presented under the same conditions. In later experiments we increased the comparability by presenting more models per experiment, as in some of the colour experiments (Fig. 73).

In this new experiment, circular discs 2, 4, 6 and 8 cm in diameter (Fig. 78 left) were used. From Fig. 80 it appears that a disc size of 4 cm is optimal.

Since we knew already at this time that the distance at which a model is presented has a great influence on its releasing value, it was very important for us to exclude this disturbing factor. It was possible to do this by moving the models very slowly (obviously subliminally) to the desired place, and then, after an interval, make it 'dance on the

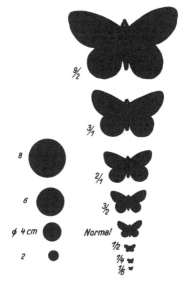

Fig. 78. Models of different sizes.

spot'. First we presented the 4 and 8 cm discs at a constant distance of 50 cm. Now we found that considerably fewer approach flights were made to the larger model than to the smaller (Fig. 81—solid lines). This was certainly due in part to the inhibitory effects of escape tendencies; this was evident not only from the fact that the largest model often released actual flight but also, in the absence of overt escape, from a lifting of the forewings, which we had learned to recognise as an incipient escape response. We concluded that in both cases the sexual behaviour was suppressed by the impulse to flee. To get round this we decided to present the models at a greater distance. When we now offered in two long series of experiments the 2, 4, 6 and

8 cm models at a distance of 100 cm the results obtained were quite different (Fig. 81—dotted line, Fig. 82). Direct observation showed no indications at all of 'fear', and these results must therefore be taken as the more reliable.

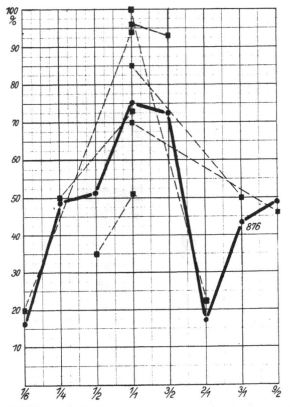

Fig. 79. The releasing value of butterfly-shaped models of various sizes with the gradual approach of the models. Thick solid line = average. Ordinate: percentage of positive responses; abscissa: size of model (see Fig. 78).

Strange to say, the natural size is thus no longer optimal, but considerably less effective than the largest one used. One might state, as before, that like the black model the largest model is 'supernormal'. Whether there is an optimal size at all, or whether it is perhaps 'the larger the better' (provided that fear responses are excluded by maintenance of a sufficiently great distance) we do not know.

These cases, in which the natural situation is suboptimal, are certainly not unique. They may well apply even to some human reactions (e.g. sexual and parental responses, eating), but in animals few well-substantiated examples seem to be on record. Koehler and Zagarus (13) who put differently coloured eggs outside the nests of brooding ringed plovers reported that a 'no doubt extraordinarily striking' white and dark spotted artificial egg was preferred over a real ringed plover egg. It appears that in the Herring Gull, as far as I can conclude from the few experiments I have done so far, the natural *size* of the egg is not optimal, for an egg model twice the normal (linear) size was selected in preference to two otherwise similar egg models of normal size.

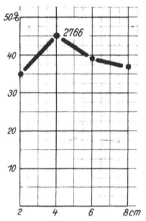

Fig. 80. The releasing value percentage of positive responses of the disc models of various sizes with the gradual approach of the models. Abscissa: size of model.

Movement Although males, as stated, also fly towards beetles, dragonflies, flies, etc. it was apparent to us that they distinctly preferred butterflies. All these insects differ considerably in size and shape from a butterfly, but since neither of these characteristics could account for the high response rate to butterflies, we presumed that the movements of the butterflies were responsible. As usual our starting point was an analysis of our own perception.

Firstly, the rather slow wingflaps appeared to us to be important; when alternately spreading and closing its wings, the butterfly changes its shape rhythmically from broad to narrow and vice versa.

Secondly, the animal as a whole moves up and down, giving the flight its typical dancing appearance. The effects of these two movements were tested separately.

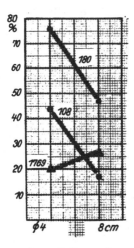

Fig. 81. The releasing value of discs of various sizes with abrupt appearance at constant distances: 50 cm distance (solid line), and 100 cm distance (broken line). Abscissa: size of model.

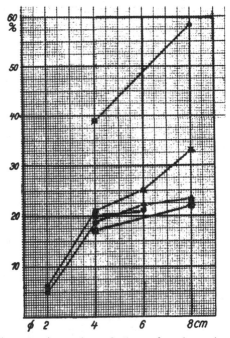

Fig. 82. The releasing value of discs of various sizes with abrupt appearance at 100 cm distance. Abscissa: size of model.

228

For this we used circular cardboard discs, 4 cm in diameter, attached to the end of a 2 m long strong wire. This model was moved in 3 ways above a sitting male; firstly as 'smoothly' as possible, showing the disc's broad side to the animal; secondly 'dancing', showing the animal the broad side as before, but raising and lowering the model irregularly over 10–30 cm in a quick rhythm (2–3 times per second); and thirdly 'turning', where the passage was smooth but the model revolved rapidly on its axis (2–3 times per second).

As Fig. 83 shows, both 'dancing' and 'turning' scored heavily. We had here certainly discovered two important characteristics. Nevertheless it appeared to us more and more, that even our best

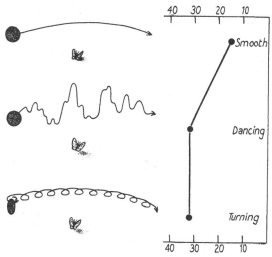

Fig. 83. The influence of movement. Abscissa: approach flight distance (i.e. the distance at which an approaching model releases an approach flight). Further explanation in text.

turning or dancing model could not compete with a real *Eumenis*. This may have been in part because we could not make this model dance and turn at the same time, which we could do perfectly well with our 'fishing rods'. Although even with this we could not compete with a real butterfly, here the difference was only minimal. We think that this, too, might have been due to the type of movement: the 'fishing rod' model moved very much more irregularly than a live butterfly. This was also suggested by the fact that, however much we tried to standardise our 'angling' movements, some of us obtained consistently higher scores than others, and differences in type of movement seemed to be the only explanation for this.

Our conclusion was confirmed by the additional fact that, of all the other insects, the one species with butterfly-like flight, *Calopteryx virgo*, elicited many more approach flight responses than any of the others (even though it was much rarer than other dragonfly species such as the Anisopterans *Sympetrum*, *Libellula*, and *Aeschna*).

Distance   While we were studying the effect of size, it occurred to us that distance might in itself be a factor. To test this, we presented the natural sized brown butterfly model as well as the butterfly models of $4\frac{1}{2}$ and $\frac{1}{6}$ times normal size. We made the model dance low above the ground and slowly approach a male whose position we had

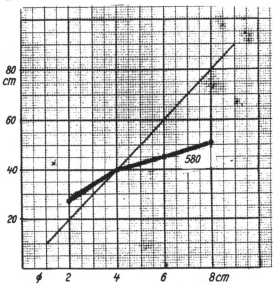

Fig. 84. Average approach flight distances for discs of various sizes obtained by slowly bringing the models nearer. Ordinate: distance of model; abscissa: size of model.

previously noted exactly. As soon as an approach flight was elicited, the model was immediately dropped on the ground and the distance between the original position of the male and the model was measured.

The natural-sized model elicited an approach flight from, on average, 82·4 cm (71 presentations), the larger model from 146 cm (23 presentations), and the small one from 19·6 cm distance (12 presentations). If the visual angle subtended by the model had been decisive, then the last two distances, extrapolated from that of the

230

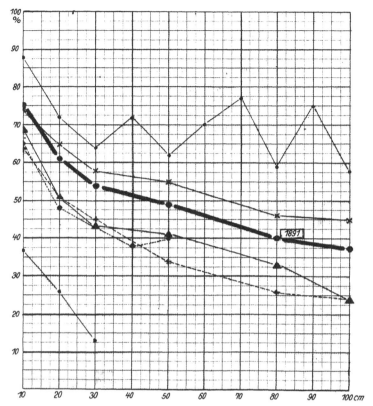

Fig. 85. The releasing value of a model when made to appear suddenly at various distances. Ordinate: percentage of positive responses; abscissa: distance. Thin curves represent individual experiments conducted on different days. Thick solid line = average.

normal model, should have been: $4.5 \times 82.4 = 371$ cm, and $\frac{1}{6} \times 82.4 = 14$ cm. Since this series of experiments was too short, we conducted a second, similar, experiment with circular paper discs 2, 4, 6 and 8 cm in diameter. The releasing values for these models have already been given in Fig. 80. Fig. 84 gives their average approach flight distance. Here we found the same discrepancy. Similar facts have been reported for ants (Homann, 11) and for *Aeschna* larvae (Baldus, 2).

From these two experiments it appears, firstly, that the butterfly has distance perception; and secondly that the model provides a stronger stimulus the nearer it is to the male; a model x times smaller than another is responded to well before it is brought x times nearer to a male.

231

To determine the effect of distance *per se*, we conducted the following experiment: we slowly (subliminally) moved a brown model of butterfly shape and standard size towards a place on the ground at a predetermined distance from a male and then, after a short pause, suddenly made it flutter on the spot. This was repeated at varying distances in a random sequence. The results, given in Fig. 85, show that the model had, in fact, a greater releasing value the nearer it was.

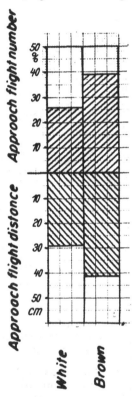

Fig. 86. Interaction between the cues of 'colour' (brightness) and 'distance'. Approach flight distance: the distance at which an approaching model releases a response. Further explanation in text.

The question as to whether this depends solely on the fact that the nearer it is to the eyes the larger it appears, will be discussed once more later (p. 239).

Interaction of Stimuli Once we knew that a model was a stronger stimulus the nearer it was, it seemed probable that in our earlier experiments, in which we had approached a male with a model until it flew at it, two models of different releasing value could have the same 'approach flight index' simply because the stronger model would, on an average, elicit a response from a greater *distance* than the weaker one. While this does not invalidate the differences found, presenting the models always at a same and constant distance would have been a more sensitive method.

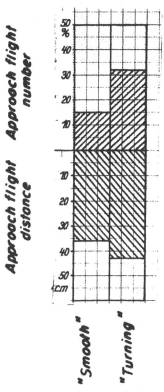

Fig. 87. Interaction between the cues of 'movement' and 'distance'.

The question as to whether 'nearness' could compensate for inadequate stimulation of other kinds was important enough to warrant another experiment. For this it would not be sufficient (as in the original method) to record each response simply as positive or negative, but we would also have to record the distance from which a model elicited an approach flight; this was done by immediately dropping the model when a male responded, and carefully remembering the exact location of the male prior to its response.

Naturally, this allowed only fairly gross estimates, especially as, since we aimed at obtaining the largest possible series, we had to keep each male in sight when it flew off after a response. We coped with this difficulty by having each 'response distance' checked by two or three people and taking the average (anyway, it was not the absolute distance but rather the ratio between the different distances that was important). First we studied the combination of distance and colour. We have seen that a white model is a much weaker stimulus than a brown one. We then worked out for each of these two models (Fig. 86) the average value of the observed response distance and found that this was considerably greater with the brown than

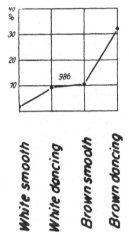

Fig. 88. Interaction between the cues of 'movement' and 'colour'.

with the white. Thus there is not only a difference in the average approach flight *number* but also in the animal's readiness to approach, for an approach flight from 42 cm (brown) indicates a greater receptiveness than one from only 29 cm (white). *Nearness can obviously compensate for suboptimal brightness.*

Incidentally, these data show clearly how 'stupidly' the animal reacts to 'characteristics'. From an anthropomorphic point of view, one might naively, but naturally expect that a suboptimal model would elicit fewer approach flights the nearer it was, since the butterfly would be able to see its shortcomings more clearly. The data show, however, that the animal simply obeys—*has to* obey—the stimulus of 'nearness'—of anything.

Before generalising, we had to check whether the same applied to other pairs of cues. We therefore also investigated the relationship between distance and movement. As Fig. 87 shows, exactly the same applies as in the case of distance-colour. In Fig. 88, finally the inter-action between 'colour' and 'movement' is shown: a dancing white model has the same stimulus value as a smoothly moving brown one.

We can therefore have confidence in the general conclusions drawn from the data that the characteristics of the models are somehow mutually pooled—another way of expressing Seitz's '*Reiz-summenregel*' ('rule of summation of stimuli').

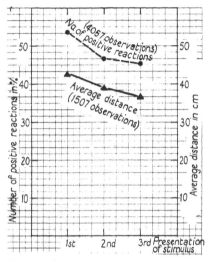

Fig. 89. The influence of repeated presentation on the approach flight number and approach flight distance.

Action-specific Waning As previously mentioned, in many experi-mental series we offered a model 3 times in succession with short pauses in between and then waited a little longer (allowing for the exchange of 'fishing rods' and recording) before the model was presented anew. This incidental peculiarity of our method made it possible for us to check whether perhaps waning occurred during a series of 3 trials.

We calculated, for some of the most regularly run series of experi-ments, the total approach flight number of all first, second, and third presentations (Fig. 89). From 1,370 first presentations, 733 (54 per cent) were positive, from 1,348 second presentations 632 (47 per cent), and from 1,339 third presentations 606 (45 per cent); so the readiness to respond falls slightly during three presentations in quick succession, and recovers in the (relatively short) intervals between 'sets-of-three'.

For some of the later experiments we were able to compare the average distance for first, second and third presentations of identical models. They show a similar drop. In a total of 1,507 positive responses, the average distance obtained for the first presentation was 43 cm, for the second 39 cm, and for the third 37 cm. Since, as we have seen, the less ready an animal is to respond, the nearer any stimulus object has to be, these figures likewise reveal a waning of the readiness to respond.

This raises the question as to whether we are dealing here with muscular fatigue or with a change at another level. Muscular ex-

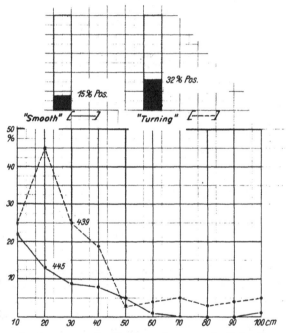

Fig. 90. The frequency distribution of approach flight distances resulting from the slow approach of a 'smooth' moving and a 'turning' model: abscissa: size of model

haustion can be excluded in view of the very short distances involved in each approach and of the species' excellent flight performance in other contexts. The waning of the readiness to respond is therefore undoubtedly behaviour-specific. How far it involves the approach flight alone, or the entire sexual behaviour, we cannot say. Since a reduced responsiveness of the eyes seems unlikely, we suspect that the change takes place in the c.n.s.

The sexual approach flight is, therefore, not 'pure response' to external stimuli but depends also on internal, more or less specific factors, though these obviously never have absolute control, since we have never observed sexual approach to nothing ('*in vacuo*'). In other words, the male's 'appetitive behaviour' is: sitting still, and being sensitive to the sight of a passing female.

Reconsideration of Some Particular Stimuli The discovery of the compensatory interaction between individual 'stimuli' showed clearly that our initial method, in which we made the models approach the animal and in which we simply counted the number of positive and negative responses irrespective of distance, tended to

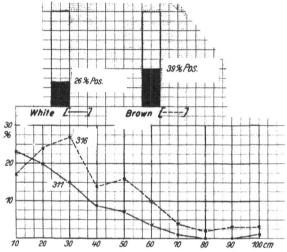

Fig. 91. The frequency distribution of approach flight distances resulting from the slow approach of a white and brown model.

obscure possible differences in releasing value of two models: because weaker models, ignored at a large distance, might still build up a high score, owing to the fact that the flight was elicited from a shorter distance than with stronger models, any differences in releasing value might not show up. (Conversely of course differences that *did* show up even in total number of positive responses must be accepted as reflecting even more pronounced differences than the mere approach index indicated, and, on the other hand, equality in both the approach index and average response distance point to real equality.) Although, in Figs. 72 and 83 the approach distances have been given, and support the approach frequency indexes, we could, with the data of those experiments in which we had measured distances, carry out a more detailed analysis. Figs. 90 and 91 show the frequency distributions of the approach distances for 'smooth' and 'turning' and for white and brown respectively. In both cases the models with the lower approach index also show a peak in the distance frequency curve that lies further to the left.

Now this in turn raises the question whether we can after all consider the data of Fig. 84 as proof of distance perception; for the smaller, weaker models we could expect a number of approaches from less than 10 cm; and the score of the larger models, is, as we have seen, lowered by the interplay of escape tendencies. However, if in Fig. 92 we compare the scores of the 2 cm and the 8 cm discs

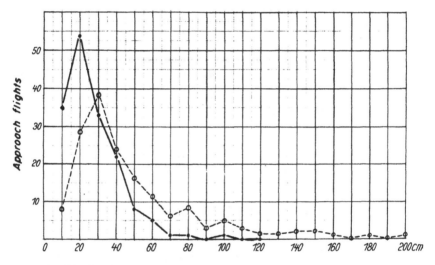

Fig. 92. The frequency distribution of approach flight distances resulting from the slow approach of an 8 cm diameter disc (dotted line) and a similar one 2 cm in diameter (solid line).

for distances at which they subtend the same angle (e.g. 10 versus 40 cm, and 20 versus 80 cm) we see even then a considerable difference in approach number.

We nevertheless decided to do a special series in which models of unequal size were made to appear suddenly at predetermined distances, now in such a way that when first seen, they subtended the same angle. Fig. 93 shows that for the pairs 4 and 8 cm (seen from 50 and 100 cm respectively) and 2 and 4 cm diameter (seen from 30 and 60 cm respectively) the number of responses is clearly different in the expected direction. The flatter slope of the 4–8 cm curve may be based on the larger model inhibiting a number of responses even when presented at 100 cm. Anyway, the combined data show clearly that the size of an object is judged irrespective of the angle subtended by it, i.e. that *Eumenis* can judge distances; 'size' and 'nearness' are genuinely different parameters.

'*Response*' *and* '*Reflex*'   While the evidence presented shows that the sexual approach flight is a response to a set of distinct visual stimuli, we are of course very far removed from an interpretation of its releasing mechanism in physiological terms. All the same, a few inferences can be made.

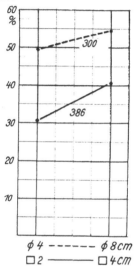

Fig. 93. The releasing value of two differently sized models which subtended the same visual angle. Further explanation in text.

(1) Our analysis of the optimal stimulus situation emphasises the configurational nature of at least some of the 'stimuli', e.g. shape (where it is obviously the *proportion* between width and length that is important) and movement (which is characterised by a time sequence of stimulation.)

(2) The phenomenon of mutual compensatory interaction between the different types of sensory input shows clearly that they are somehow, inside the animal, 'pooled'. Even though we have no morphological evidence for the existence of a 'centre', this pooling effect forces us to conclude that the nervous system contains a centre, *in the functional sense*. Since the response is always the same, namely

coordinated flight, there is no question of particular stimuli feeding directly into particular parts of the muscle complex involved. The difference between the two functional types is shown diagrammatically in Fig. 94.

(3) The output of the system is a complex motor pattern, in which not only eliciting but also orienting stimuli must play a part. Our experiments give no evidence on the nature of these orienting stimuli, but even so the alternation between up-and-down stroke of the wings, and the left–right synchronisation show how complex this orientation must be.

All this highlights the difference between this 'response' and the conventional 'reflex'. The methodological differences between the physiologist who isolates, in the laboratory, the simplest possible reflexes and the behaviourist who analyses the complex stimulus situations controlling a complex response makes it difficult to judge to what extent the two phenomena really differ. One suspects that the term 'reflex' is often very loosely applied even to the simpler cases; thus the well-known 'wiping reflex' of the frog undoubtedly contains an orientation component and even without this forms a sequence. But it is clear that the sexual approach of the Grayling male is neither a simple reflex nor a parallel bundle of reflexes. We emphasise this because for instance Doflein (6) and Szymanski (20) have applied the term 'reflex' to movements in the ant lion larva and in the snail *Helix* respectively, which may well, if analysed in the same way as we have done with the sexual approach flight of the male Grayling, turn out to be similarly complex. Little is gained by thus loosely applying the term 'reflex' to behaviour patterns, and suggesting an identity that may well be spurious. For the moment the term 'reflex' seems therefore to be of little use in behaviour studies.

The Function of the Scent Areas The Grayling male possesses a particularly striking secondary sexual characteristic: a patch of dark scales on the upper side of its forewings. This dark patch, together with the brownish tinge to the yellow areas, give the upper surface of the male a much duller brown colour than the female.

The dark patch appears, under low magnification, to be a special structure. Between the normal wide and flat scales which are found over the entire wing, there are in this area a large number of very small and slender scales which carry a fine branching brush-like structure at their distal end (Fig. 95).

Similar structures can be found in various other species. Figs 96 and 97 show the dark patches and their scales in *Epinephele jurtina*, *E. tithonus* and *Pararge megera*.

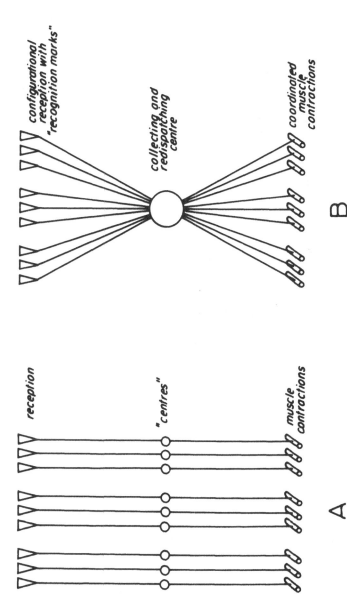

Fig. 94. Symbolic representation of a group of simple reflex actions (A) and a 'response' (B). The centres are represented by circles.

The structure of these brush-carrying scales reminds one of the so-called 'scent scales' already found in many species. Fritz Müller (**18**) was the first to infer that such brush-like scales might facilitate the evaporation of a volatile substance. He found that certain Brazilian butterflies, especially those bearing brush-shaped scales, produce a strong, often pleasing perfume, noticeable even to humans, and different from one species to another. Aurivillius (**1**) thought that these scent scales were only found in the males and named them, therefore, 'androconia'.

Their structure appears to be specially modified for the evaporation of scent; the ramifications of the distal end enormously increase the surface area for evaporation; for some species it is also known that scent is produced by glandular cells which lie at the base of the wing scales (Freiling, **8**).

Fig. 95. Two ordinary scales and one scent scale from *Eumenis semele*.

In *Eumenis semele* we likewise found cells at the base of the scent scales comparable to those found by Freiling in *Apodaea lineata*; a study of these structures is in progress.[1]

Evidence for the suggested sexual significance of such scent scales is sparse. It is true that there are many observations that suggest such a function but experimental evidence seems to be lacking. Deegener (**4**) has shown that the males of *Hepiales hecta* release their scent during their peculiar courtship flight; this may lure the females to them. In other species this chemical luring is practised by the females; Dickens (**5**) has evidence of this in *Ephestia kühniella* (Zeller) and its relatives *E. elutella* (Hb.); *E. cautella* (Hb.) and *Plodia interpunctella* (Hb.).

Meanwhile Fritz Müller was aware that the function of the androconia could not always be that of attracting the female from a distance, since the males seek out the females and not the other way round.

---

[1] (Author's note) Yet has never been concluded.

Fig. 96. Dorsal views of males (top) and females of (left to right): *Epinephele jurtina* L; *E. tithonus* L; *Pararge megera* L. Right-wing scent fields outlines in black.

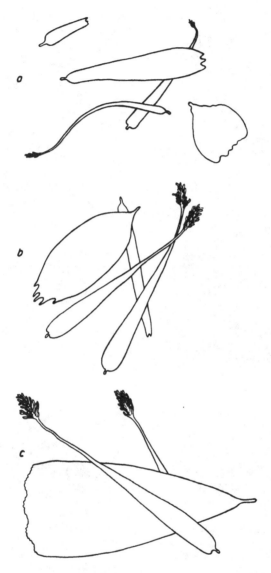

Fig. 97. Scent scales and ordinary scales of: a. *Epinephele jurtina*; b. *E. tithonus*; c. *Pararge megera* L.

From our account it will be clear that, in *Eumenis*, too the androconia are unlikely to attract the females, for here too it is the males which seek out the females. The later stages of courtship suggest rather that they have a stimulating function during the actual courtship. Fanning, that is the flapping of the leading edges of the wings while their remaining parts stay closed, must result in a pulsating air current; each time the male's wings are closed the female must be hit by an air puff emanating from their frontal half, thus from the scent areas. Since in wing quivering too, at least initially, the leading edges of the wings are slightly opened, the same is probably true here. In bowing, the role of the scent areas may be even more marked since at least one, but usually both, of the female's antennae are captured between the male's forewings. The antennae doubtless carry the scent receptors. Uncompleted studies on the structure of the antennal sense organs in *E. semele* indicated that the olfactory organs are, in fact, most numerous on the tips of the antennal knobs. This has, to our knowledge, actually been demonstrated for one species of the Rhopalocera; Knoll (**14**) found that the response to scent substance in *Charaxes jasius* (L) disappeared after amputation of the antennal knobs.

During the bowing of the male, the *E. semele* female bends the antennal knobs outwards in such a manner that the terminal surface of the knobs lie almost parallel to the upper surface of the male's wings. The male's fanning, bowing and the bent position of the female's antennae are thus all behaviours that would fit in well with our interpretation of the courtship.

Females that are not completely willing to mate show wing flapping, which has a repelling function, and if the male persists, will jump or fly away at exactly the instant the male presses its wings together in bowing. Thus it is at this instant that the decision of the female is made, and this too shows that the closing of the wings exerts a strong stimulus upon the female.

That the androconial area helps to bring the female through courtship to pairing was demonstrated as follows. We placed a number of females in our flight cage together with males whose scent areas had been removed; and with males whose scent areas were left intact but of whom a neighbouring part of the wings was treated in the same way as in the first group. A stimulating effect of the scent areas would be evident if the females accepted males of the second group more readily than the 'de-scented' males. The experiments were done mainly in July when virgin females were obtainable in sufficient numbers. On fine, sunny mornings we usually found several newly emerged females, and even more males. By midday we had treated half the males in the following manner: the androconial area was carefully brushed clean with a fine, dry brush. We then covered the cleared portion of the wings with a thin layer of shellac dissolved in alcohol. When the shellac had completely dried, we marked the animal with a number on the undersides of both hindwings, using aluminium paint. The animal was then released into the flight cage. These males were termed 'lacquered males'.

The second group of males, the 'control males', were treated in exactly the same way, only in their case we did not lacquer the scent areas but a nearby area of the upper forewing surface.

After these males had also been released into the flight cage, the females were introduced and observations began.

Since it was difficult to keep the animals fit for more than one day, we released all animals after each experiment and replaced them next day with new ones. Since courtship took place in the early afternoon, our observations were limited to only a few hours per day, and the animals were willing to mate only under sunny and windless conditions. Thus in our climate only a few days were suitable for these experiments and only very few were possible at all.

Before we could conclude anything from our observations, we had to ensure that our operations on the scent areas did not reduce the males' readiness to mate. It was immediately obvious that if there were any difference at all between the two groups, it was rather to the advantage of the 'lacquered males'. Most noticeably in the 'lacquered' male group there were animals which continually bothered the females: as anticipated, they were not accepted by the females and because of this continued their mating attempts for a long time, while the 'control' males, when ready to copulate, soon reached their goal.

In the first summer, when the 'control' males had lacquered hindwings a total of 9 control males and 5 lacquered males reached copulation. In the next summer, when the control males also had a portion of the forewings lacquered, these figures were 7 and 6 respectively.

These numbers can scarcely be taken as proof of our hypothesis. Nevertheless, this experiment did yield some qualitative evidence, since the copulations of the lacquered males were often arrived at in an irregular manner. In one case, for instance, we saw a female being courted by a control male, when suddenly a lacquered male appeared and at the last moment rushed in and 'stole' a mating. These and similar irregularities always occurred in the upper corners of the flight cage where many butterflies were often gathered together. In such clusters, which contained more animals than we had ever seen together in the wild, courting pairs were often disturbed. Thus if the scent stimulus plays a role, the difference between lacquered males and control males in clusters was probably obscured; it appeared that the success of the lacquered males was enhanced by the presence of the control males. To test this, and the same time to establish a generally more suitable experimental set-up, we used, in the third summer, two flight cages, one for each category of males. The group of lacquered males was, in each experiment, as large as that of the control males; the females caught were divided into two equal groups. One group of females was released into the flight cage with the lacquered males and the other into that containing control males. On one day one flight cage contained the lacquered males, on the next, the control males.

The results were now 9 to 2 in favour of the control males. This gave a total over the three years of 25 copulations for the control males and 13 for the lacquered males. The 'stolen' matings are included in these 13 copulations, since we could not accurately determine their number.

These experiments, although few in number, support the theory that the androconial area in the male is an important means of stimulating the female. The male's courtship appears to be well adapted to stimulating the female by the scent from the androconial areas, preventing both avoidance and repelling behaviour.

That visual stimuli play at best a minor part is indicated not only by the form of the courtship but also by the fact that the brightly coloured upper surfaces of the males are not (as in species that use visual courtship) displayed clearly to the partner. Furthermore, in 1941 a female that had been blinded for other purposes and released in the field, copulated readily with a male. She did not simply permit copulation, but after the bowing of the male, lifted her wings. Finally, it is difficult to believe in visual stimulation once one has seen how drastically the male's appearance is changed by the large glittering aluminium number. A male marked with a '30', for example, is hardly recognisable as an *Eumenis*, yet such a control male can arrive at copulation.

Summary

Most of this work was concerned with the analysis of the courtship behaviour of *Eumenis semele* (L). The first response of a male to a female flying past is very unselective. The effects of differences in colour, shape, size, distance and movement were investigated.

The female's colour is not important; the patterning of the wings has no effect at all, but the darkest female (model) elicits most approach flights. Black and red have a higher releasing value than the

247

normal coloration; they are 'supernormal'. Yet *E. semele* is anything but 'colourblind'; feeding animals respond to yellow and blue flowers or paper models; this is a genuine response to wavelength. Thus the sensitivity to stimuli varies with the motivation.

The female's shape is of very little importance: butterfly shapes, discs and rectangles are flown towards with about equal frequency. Rectangles of the same surface area but with varying proportions of length to breadth do not all have the same releasing value; here a relationship rather than the absolute size of the surface area is effective.

A model of natural size is more effective than a smaller model, but inferior to larger models. Here too 'supernormal' stimulation was possible.

All models are more effective the nearer they are. This is not due solely to the size of the retinal image, but also to genuine distance perception, which however was not analysed further.

Movement, and the type of movement, are very important: a smoothly moving model has a far lower releasing value than a 'dancing' one or one which, by turning continually, changes its appearance from broad to narrow.

The interaction of these characteristics is mutually reinforcing; a deficiency in one characteristic can be compensated for by any of the other ones. There thus appears to be a 'pooling station' between receptors and effectors in which the dissimilar inputs produced by the separate stimuli are united in some way and transmitted as a whole. The sexual approach flight seems far removed from a 'simple reflex'.

The function of the 'scent areas' on the male's forewings was studied further by extirpation experiments. The androconia obviously release a scent which makes the female accept a male.

The movements of the male's wings and the female's antennae during courtship are such that the male's scent stimulus has the greatest opportunity of being carried to the female's chemoreceptors.

REFERENCES

1 AURIVILLIUS, CH. (1880). 'Über sekundäre Geschlechtscharaktere nordischer Tagfalter', *Bib. Svenska Vet. Akad. Handl.*, 5 (quoted from DEEGENER).
2 BALDUS, K. (1926). 'Experimentelle Untersuchungen über die Entfernungslokalisation von Libellen', *Z. vergl. Physiol.*, 3, 373–513.
3a BUDDENBROCK, W. VON (1929). 'Einige Bemerkungen zum augenblicklichen Stand der Frage nach dem Farbensinn der Tiere', *Zool. Anz.*, 84, 189–201.
3b —— (1937). *Ders. Grundriß der vergl. Physiologie*, 2. Aufl., 1, Bd., Berlin.
4 DEEGENER, P. (1902). 'Das Duftorgan von *Hepiales hectus* (L)', *Z. wiss. Zool.*, 71, 276–96.

5 DICKENS, G. R. (1930). 'The scent glands of certain Phycitidae (Lepidoptera)', *Trans. Roy. Entomol. Soc. London*, **85**, 331–62.

6 DOFLEIN, FR. (1916). *Der Ameisenlöwe*, Jena.

7 ECKSTEIN, K. (1913). *Die Schmetterlinge Deutschlands*, Stuttgart.

8 FREILING, H. H. (1909). 'Duftorgane der weiblichen Schmetterlinge usw', *Z. wiss. Zool.*, **92**, 210–91.

9 FROHAWK, F. W. (1914). *The Natural History of British Butterflies*, London.

10 HEIMANS, E. and JAC. P. THIJSSE (1913). *Hei en Dennen*, Amsterdam.

11 HOMANN, H. (1924). 'Zum Problem der Ocellenfunktion bei den Insecten', *Z. vergl. Physiol.*, **1**, 541–78.

12 ILSE, D. (1928). 'Über den Farbensinn der Tagfalter', *Z. vergl. Physiol.*, **8**, 658–92.

13 KOEHLER, O. and A. ZAGARUS (1937). 'Beiträge zum Brutverhalten des Halsbandregenpfeifers (*Charadrius h. hiaticula* L.) Beitr. Fortpflanzungsbiol', *Vögel*, **13**, 1–9.

14 KNOLL, FR. (1926). *Insekten und Blumen*, Vienna.

15 LORENZ, K. (1927). 'Beobachtungen an Dohlen, *J. f. Ornithol.*, **75**, 511–19.

16 LOTMAR, R. (1933). 'Neue Untersuchungen über den Farbensinn der Bienen, mit besonderer Berücksichtigung des Ultravioletts', *Z. vergl. Physiol.*, **19**, 673–724.

17 MINNICH, D. E. (1922). 'The chemical sense of the tarsi of the Red Admiral butterfly', *J. Expt. Zool.*, **35**, 57–82.

18 MÜLLER, F. (1877). 'Über Haarpinsel, Filzflecke und ähnliche Gebilde auf den Flügeln männlicher Schmetterlinge', *Jenaische Z. Naturw.*, **11**, 99–115.

19 SCHLIEPER, C. (1926). 'Der Farbensinn von *Hippolyte*, zugleich ein Beitrag zum Bewegungssehen der Krebse', *Verh. D. Zool. Gesellsch.*, **31**, 188–93.

20 SZIMANSKY, J. S. (1913). 'Ein Versuch, die für das Liebesspiel charakteristischen Körperbewegungen bei der Weinbergschnecke künstlich hervorzurufen', *Arch. ges. Physiol.*, **149**, 471–82.

21 TINBERGEN, N. (1941) 'Ethologische Beobachtungen am Samtfalter *Satyrus semele* (Vorl. Mitt.)', *Journal f. Ornithologie*, Erg.-Bd. III (Festschrift für O. HEINROTH) S. 132–44.

22 WEIS, I. (1930). 'Versuche über die Geschmacksrezeption durch die Tarsen des Admirals *Pyrameis atalanta* L.', *Z. vergl. Physiol.*, **12**, 206–48.

(From the Department of Zoology, Oxford University)

6
Egg Shell Removal by the Black-headed Gull, *Larus ridibundus* L.

A Behaviour Component of Camouflage (1961)

I. Introduction

Many birds dispose in one way or another of the empty egg shell after the chick has hatched. A shell may be built in or trampled down; it may be broken up and eaten; or, more usually, it is picked up, carried away and dropped at some distance from the nest. C. and D. Nethersole Thompson (11), who have given a detailed summary of our knowledge of egg shell disposal in birds, emphasise the inter- and even intraspecific variability of the responses involved. Since, in addition, the actual response is often over in a few seconds, and happens only once or twice for each egg, it is not surprising that our knowledge is still fragmentary. On the whole, the presence or absence of the response and its particular form seems to be typical of species or groups of species; for instance, it seems to be absent or nearly so in Anseres and in Gallinaceous birds; Accipitres often break up and eat the shell; Snipe are said to be 'particularly lax' (Nethersole Thompson); Avocets, *Recurvirostra avosetta* L., remove discarded egg shells anywhere in the colony (Makkink, 9). In the many species which carry the egg shell away, the response, occurring as it does just after hatching, when the young birds need warmth and protection from predators, must be supposed to have considerable survival value.

The Black-headed Gull invariably removes the egg shell in a matter of hours after hatching (Fig. 98); it is extremely rare to find an egg shell in the nest once the chicks have dried. We have only a few direct observations on the time lapse between hatching and carrying in undisturbed gulls, but the 10 records we have (1', 1', 15', 55', 60', 105', 109', 192', 206', 225') suggest that the response is usually not very prompt. The carrying is done by the parent actually engaged at the nest, never (as far as we know) by the non-brooding partner which may be standing on the territory, even when it stands next to

250

the sitting bird. At nest relief either the leaving partner, or, more often, the reliever carries the shell. Often however it is the sitting bird who starts looking at the shell, stretches its neck towards it, takes it in its bill and nibbles it (sometimes breaking off fragments while doing so, which then are swallowed), and finally rises and then either walks or flies away with the shell in its bill. The shell is dropped anywhere between a few inches and a hundred yards from the nest. We have also observed birds which flew off with the shell, made a wide loop in the air, and descended again at the nest with the shell still in their bill which they then either dropped on the nest or carried effectively straight away. There is no special place to which the shell is carried, though there may be a slight tendency to fly against the wind, or over an updraught, or where the carrier is less likely to be harrassed by other gulls; almost always the shell lands well beyond the

Fig. 98. The first stages of egg shell removal (left) and egg retrieving (right) in the Black-headed Gull, *Larus ridibundus* (see below).

territory's boundary. On rare occasions a shell may land in a neighbour's nest—where the latter then treats it as one of its own shells, i.e. removes it.

II. Statement of the Problems

During our studies of the biology and the behaviour of gulls this response gradually began to intrigue us for a variety of reasons. (1) The shell does not differ strikingly from an egg, since it is only the small 'lid' at the obtuse end which comes off during hatching; yet it is treated very differently from an egg, and eggs are never carried away. This raised the question of the stimuli by which the gulls recognise the shell. Systematic tests with egg shell dummies could provide the answer. (2) What could be the survival value of the response? (a) Would the sharp edges of the shell be likely to injure the chicks? Nethersole Thompson (**11**) raises this possibility, adding that poultry breeders know this danger well. (b) Would the shell tend to slip over an unhatched egg, thus trapping the chick in a double

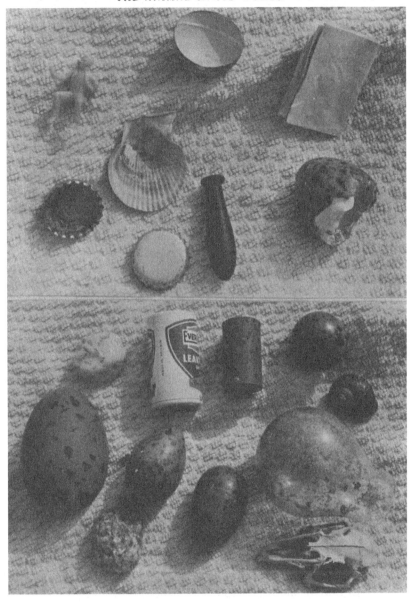

Fig. 99. Variety of objects that elicit egg shell removal (top) and incubation (bottom) by Black-headed Gulls.

shell? (c) Would the shells interfere in some way with brooding? (d) Would the moist organic material left behind in the shell provide a breeding ground for bacteria or moulds? (e) Would egg shells, if left near the nest, perhaps attract the attention of predators and so endanger the brood?

The following facts, obtained earlier by our co-workers, seemed to give some clues:

(a) C. Beer (3) found that Black-headed Gulls do not merely carry shells but a great variety of objects as well if they happen to be found in the nest. Some of these objects are shown in Fig. 99. It seemed that the best characterisation of this class of objects would be: 'Any object—perhaps below a certain size—which does not resemble an egg, or a chick, or nest material, or food'; in short: 'any strange object'. The very wide range of objects responded to suggests that the birds react to very few sign stimuli; it might be that the response was adapted.to deal with a much wider range of objects than just the egg shell.

(b) C. Beer (3), testing the gulls' readiness to show his response at different times of the season, offered standard egg shell dummies (halved ping-pong balls painted egg shell colour outside, Fig. 99) to a large number of gulls once every day from the moment nest scrapes were formed (which is up to about 3 weeks before the laying of the first egg) till well beyond the hatching of the chicks. He found that under these conditions of standard (and near-optimal) stimulation the response could be elicited from at least 20 days before laying till 3 weeks after hatching. In this respect the response behaves rather like typical incubation responses such as sitting and egg retrieving which also develop gradually in the pre-egg period (Beer, 3). In view of the heavy predation to which eggs are subjected (see below), and of the fact that the eggs are otherwise carefully guarded, this fact suggests that the response is important throughout the incubation period, and not merely during the few days when the chicks hatch.

(c) Finally, E. Cullen (4) found that the Kittiwake, *Rissa tridactyla* L., never carries the egg shell. The shells are just left in the nest until they are accidentally kicked off. It is true that this often happens in the first few days after hatching, but shells occasionally stay in the nest or on the rim for weeks, and at any rate they remain in or on the nest much longer than is the case with the Black-headed Gull. The Sandwich Tern, *Sterna sandvicensis* Lath., does not remove the egg shells either (J. M. Cullen, 5).

These observations combined suggest that neither the avoidance of injury, nor of parasitic infection, nor of interference with brooding are the main functions of egg shell removal—if this were so, then the

Kittiwake as the most nidicolous species of gull would not lack the response. The most likely function seemed to be the maintenance of the camouflage of the brood—neither Kittiwake nor Sandwich Tern can be said to go in for camouflage to the extent of the other gulls and terns.

Thus these observations naturally led to an investigation into the function of the response and to a study of the stimuli eliciting it. In the following we shall deal with the problem of survival value first.

III. The Survival Value of Egg Shell Removal

The assumption that egg shell removal would serve to maintain the camouflage of the brood presupposes that the brood is protected by camouflage. This basic assumption, usually taken for granted but—as far as we know—never really tested, was investigated in the following way.

First, we collected whatever observations we could about predation in the colony. While these observations are largely qualitative, they show convincingly that predation is severe throughout the season.

Very many eggs disappear in the course of the spring. We did not make systematic counts but can give the following qualitative data. In the Ravenglass colony Carrion Crows, *Corvus corone* L., did not account for many egg losses, because, as we could observe time and again, they are easily chased away by the mass attack of the Black-headed Gulls. In fact we never saw a Crow alight in the colony. However, attacks by one, two or three gulls (which often occurred in our tests) did not deter Crows, and we must assume that nests on the fringe of the colony, and the dozens of nests we regularly find outside the colony and which do not survive, often fall victims to the Crows.

Egg predation within the colony was due to the following predators. Three pairs of Herring Gulls, *Larus argentatus* Pont., which bred in the gullery levied a constant toll of eggs, and later of chicks. Although the Black-headed Gulls attacked them, they could not altogether stop them from snatching eggs and chicks. We observed hundreds of occasions on which non-breeding Herring Gulls and Lesser Black-backed Gulls, *Larus fuscus* L. (many of them immature), passed near or over the colony or over our tests. On only one occasion did we see any of these taking an egg; they usually were totally uninterested in the colony.

Black-headed Gulls prey on each other's eggs to a certain extent. Most of those who visited our experiments did not attack an un-damaged egg (the few exceptions are mentioned in our tables) but finished an egg once it had been broken by other predators. Later in the season individual Black-headed Gulls specialised on a diet of newly hatched chicks (see section VI).

Foxes, *Vulpes vulpes* L., which regularly visited the colony, and killed large numbers of adults early in the season and many half or fully grown chicks towards the end, did visit the gullery in the egg season, but we have no direct evidence of the amount of damage done by them in this part of the year. The gulls were greatly disturbed whenever a Fox entered the colony, but they did not attack him as fiercely as they attacked Herring Gulls and Crows. They flew over the Fox in a dense flock, calling the alarm, and made occasional swoops at it.

Stoats, *Mustela erminea* L., and Hedgehogs, *Erinaceus europaeus* L., visited the gullery and probably accounted for some losses. The gulls hovered over them in a low, dense flock but did not quite succeed in deterring them. In spite of several thousands of man-hours spent in hides in the gullery by us and by our colleagues Dr C. Beer and Dr G. Manley we never saw either a Fox, a Stoat or a hedgehog actually taking an egg; they stayed in the dense cover, and though we have been able to read a great deal from their tracks whenever they had moved over bare sand, this method is of no avail in the egg season since the gull's nests are situated in vegetation.

Surprisingly, the Peregrine Falcons, *Falco peregrinus* L., which were often seen near or over the gulleries, left the Black-headed Gulls in peace, though they regularly took waders and carrier pigeons. The gulls panicked however when a Peregrine flew past.

In order to eliminate the effect of the gulls' social nest defence we put out eggs, singly and widely scattered (20 yds apart) in two wide valleys outside the gullery proper. These valleys had a close vegetation of grasses, sedges and other plants not exceeding 10 cm in height. Each egg was laid out in a small depression roughly the size of a Black-headed Gull's nest. Two categories of eggs were arranged in alternate pattern, and in successive tests exchanged position. While predators, particularly Carrion Crows, showed remarkable and quick conditioning to the general areas where we presented eggs, there was no indication that the exact spots where eggs had been found were remembered, and in any case such retention, if it would occur, would tend to reduce rather than enhance differential predation. We assumed (erroneously) that it might well be days before the first predator would discover the eggs, and in our first tests we therefore did not keep a continuous watch. We soon discovered however that the first eggs were taken within one or a few hours after laying out the test, and from then on we usually watched the test area from a hide put up in a commanding position, allowing a view of the entire valley.

The tests were usually broken off as soon as approximately half the eggs had been taken.

Experiment 1 Not wishing to take eggs of the Black-headed Gulls themselves (the colony is protected) we did our first test with hens' eggs, half of which were painted a matt white, the other half painted roughly like gull's eggs. To the human eye the latter, though not quite similar to gulls' eggs, were relatively well camouflaged. It soon became clear that we had underrated the eye-sight of the predators, for the 'artificially camouflaged' eggs were readily found. In 9 sessions, lasting from 20 min to $7\frac{1}{2}$ hr, we saw Carrion Crows and Herring Gulls take the numbers of eggs (out of a total presented of 104) given in Table 12.

Table 12. *Numbers of artificially camouflaged hens' eggs ('Artif. cam') and white hens' eggs ('White') taken and not taken by predators*

| | Artif. cam. | | White | |
	taken	not taken	taken	not taken
Carrion Crows	16		18	
Herring Gulls	0		1 (+3)	
	—			
	16	36	19 (+3)	33 (−3)

No. of presentations: 2×52. Difference between 'artif. cam.' and 'white' not significant.[1]

[1] The p-Values given in the tables were calculated by the χ^2-method, except where stated otherwise.

We are astonished to see how easily particularly the Carrion Crows found even our 'camouflaged' eggs. Each test area was usually visited every few hours by a pair of Crows.[1] They would fly in at a height of about 6 m, looking down. Their sudden stalling and subsequent alighting near an egg were unmistakable signs that they had seen it; they usually discovered even camouflaged eggs from well over 10 m distance. Often the Crows would discover every single egg, whether white or camouflaged, in the area over which they happened to fly. They either carried an egg away in their bills without damaging it to eat it elsewhere, or opened it on the spot and ate part of the contents, or (usually after they had first eaten 4 or 5 eggs) they carried it away and buried it. In some cases we saw Crows uncover these buried eggs one or more days after they had been cached.

Of the numerous Herring Gulls and Lesser Black-backed Gulls

[1] We have good reasons to believe that each of our two test areas were visited by one pair.

living in the general area, many of which flew over our test area every day, interest in eggs was shown only by three resident pairs of Herring Gulls. Their eyesight was undoubtedly less keen than that of the Crows (as expressed in the scores in Table 12 and particularly Table 13); further they were remarkably timid. For instance in experiment 1 we were certain on three separate occasions that a Herring Gull had discovered a white egg but did not dare approach it (the '3' in parentheses in Table 12 refers to these occasions). Yet as soon as Crows began to search the area, Herring Gulls would appear and attack them. Often Crows and Gulls attacked each other mutually, swooping down on their opponents from the air, for minutes before either of them alighted near an egg. Usually the Gulls succeeded in claiming an egg first. There were also occasions on which the Crows had the area to themselves.

It seemed obvious that our 'camouflage' was not effective at all. Now the artificially camouflaged eggs differed from real gulls' eggs in four respects: (1) the ground colour, though to the human eye matching the overall colour of the rather brownish background very well, was slightly different from the ground colour of most of the gulls' eggs; (2) the dark grey dots which we painted on the eggs were more uniform in size and distribution than those on the real gulls' eggs; (3) unlike the natural dots they were all of one hue; and (4) hens' eggs are considerably larger than Black-headed Gulls' eggs, and hence probably more conspicuous.

Experiment 2   Therefore, we next tested real, unchanged Black-headed Gulls' eggs against Black-headed Gulls' eggs painted a matt white. The tests were conducted in the same way as the previous ones. The results, obtained in 12 sessions lasting from 20 min to 4 hr, in which 137 eggs were presented, are summarised in Table 13.

Table 13. *Numbers of normal Black-headed Gulls' eggs ('Natural') and Black-headed Gulls' eggs painted white ('White') taken and not taken by predators*

| | Natural | | White | |
	taken	not taken	taken	not taken
Carrion Crows	8		14	
Herring Gulls	1		19	
Black-headed Gulls	2		7	
Unknown	2		3	
	—		—	
	13	55	43	26

No. of presentations: 68 + 69. Difference between 'natural' and 'white' significant at 0·1% level.

257

Experiment 3 In order to get an impression of the parts played by size and by the nature of our 'artificial camouflage', we painted Black-headed Gulls' eggs in the same way as the 'camouflaged' hens' eggs, and compared their vulnerability with that of Black-headed Gulls' eggs painted white. The results of this test, to which we devoted only 5 sessions of from 1 to 3 hr, and in which 48 eggs were presented, are given in Table 14.

Table 14. *Numbers of artificially camouflaged Black-headed Gulls' eggs ('Artif. cam.') and Black-headed Gulls' eggs painted white ('White') taken and not taken by predators*

| | Artif. cam. | | White | |
	taken	not taken	taken	not taken
Carrion Crows	4		5	
Herring Gulls	4		8	
Black-headed Gulls	1		1	
	—		—	
	9	15	14	10

No. of presentations: 24+24. Difference between 'artif. cam.' and 'white' not significant ($20\% < p < 30\%$).

From these experiments, and particularly from experiment 2, we conclude that the natural egg colour of the Black-headed Gulls' eggs makes them less vulnerable to attack by predators hunting by sight than they would be if they were white; in other words that their colour acts as camouflage. The difference between the results of experiments 1 and 3 on the one hand and experiment 2 on the other indicates that we had underrated the eyesight of the predators, and also that their reactions to the different aspects of camouflage deserve a closer study: the parts played by over-all colour, by pattern and hue of the dotting, perhaps even by the texture of the eggs' surface we hope to investigate later—it was a pleasant surprise to discover that large-scale experiments are possible. For our present purpose we consider it sufficient to know that painting the eggs white makes them more vulnerable.

Experiment 4 We can now turn to the question whether or not the presence of an egg shell endangers an egg or a chick nearby. For obvious reasons we chose to investigate this for eggs rather than for chicks. The principle of this experiment was the same as that of the previous ones. We laid out, again avoiding site-conditioning in the predators, equal numbers of single Black-headed Gulls' eggs with and without an empty egg shell beside them. The shells used were such from which a chick had actually hatched the year before and which we had dried in the shade and kept in closed tins. The shells were put at about 5 cm from the eggs which were again put in nest-shaped pits. Predators were watched during 5 sessions lasting from 45 min to 4½ hr. Sixty eggs were used. The results are given in Table 15.

Table 15. *Numbers of Black-headed Gulls' 'Eggs with shell' and 'Eggs without shell' taken and not taken by predators. Eggs not concealed*

	Eggs with shell		Eggs without shell	
	taken	not taken	taken	not taken
Carrion Crows	6		7	
Herring Gulls	9		7	
	15	15	14	16

No. of presentations: 30 + 30. Difference between 'eggs with shell' and 'eggs without shell' not significant.

We did not consider this result conclusive because the circumstances of the experiment differed from the natural situation in two respects. (1) Although a nest in which a chick has recently hatched may contain unhatched eggs, there are chicks equally or rather more often; and chicks, apart from having a less conspicuous shape than eggs, do at a quite early stage show a tendency to crouch at least half concealed in the vegetation when the parent gulls call the alarm. The Crows and Gulls might have less difficulty finding eggs than seeing chicks. (2) Both predators were always vigorously attacked by many Black-headed Gulls whenever they came in or near the gullery. In avoiding these attacks their attention is taken up for the greater part of the time (as judged by their head movements and evading action) by keeping an eye on the attackers; in the natural situation they never have the opportunity to look and search at their leisure. And to predators searching for camouflaged prey leisure means time for random scanning and opportunity for undivided attention—both probably factors enhancing discovery, i.e. fixation of non-conspicuous objects. In other words our experiment had probably made things too easy for the predators, even though in the colony the nests themselves are often visible from a distance.

Experiment 5 We therefore decided to repeat this experiment with slightly concealed eggs. This was done by covering each egg (whether or not accompanied by an egg shell) with two or three straws of dead Marram Grass, a very slight change which nevertheless made the situation far more similar to that offered by crouching chicks. Most tests of this experiment were done without watching from a hide; we knew by now who the main predators were, and for our main problem it was not really relevant to know the agent. In 8 tests, lasting from 2 hr 40 min to 4 hr 40 min, 120 eggs were offered, of which 60 with a shell at 5 cm distance. The results are given in Table 16.

Table 16. *Numbers of Black-headed Gulls' 'Eggs without shell' taken and not taken by predators. Eggs slightly concealed*

Eggs with shell		Eggs without shell	
taken	not taken	taken	not taken
39	21	13	47

No. of presentations: 60 + 60. Difference between 'eggs with shell' and 'eggs without shell': $p < 0.1\%$.

The conclusion must be that the near presence of an egg shell helps Carrion Crows and Herring Gulls in finding a more or less concealed, camouflaged prey, and that therefore egg shells would endanger the brood if they were not carried away.

Experiment 6   This was a series of pilot tests designed to examine whether the Carrion Crows would be readily conditioned to shells once they had found eggs near shells, and whether, if they would find shells without eggs, they would lose interest in the shells. These tests began on April 21, 1960, after the Crows operating in the valley had already gained experience with white and camouflaged eggs, and, since they had broken many of those while eating them, could have learned that egg shells meant food. In the very first test, egg shells were laid out in the western area of the valley, where in previous tests with whole eggs the Crows had never been seen to alight. At their first visit the Crows alighted near these shells, pecked at them, and searched in the neighbourhood. They left after a few minutes whereupon we took the shells away. A few hours later we again laid out egg shells without eggs, this time scattered over the whole valley. When the Crows next returned they flew round over the valley, looking down as usual, but they did not alight. Next morning semi-concealed eggs were laid out over the whole valley, each with an egg shell at the usual distance of 5 cm. This time the Crows did alight and took a number of the eggs. A few hours later we once more laid out shells only, scattered over the valley, and at the next visit the Crows came down near several of them and searched. The morning after this we laid out just shells, this time in the eastern part of the valley. This time the Crows did not alight. When next, between this and their next visit we gave each shell an egg at 5 cm, the Crows alighted again when they returned and took several eggs. Later that same day just shells were given in the western part, which could not induce the Crows to alight. Such tests were continued until April 30, and while they did not give sufficient information, they strongly suggest: (a) that one experience with shells-without-eggs was sufficient to keep the Crows from alighting near the shells the next time; and (b) that renewed presentation of eggs with the shells attracted them again. Further, we had the impression (c) that the Crows later learned that egg shells in a part of the valley where shells only had been presented several times meant no food—in other words that they associated 'shells only' with the locality. The full record of these experiments is given in Table 17.

Table 17. *Record of 'conditioning test' explained in text*

Es+ = eggs+shells offered, eggs taken; Es− = eggs+shells offered, eggs not taken; S+ = shells alone offered, crows alighted and searched; S− = shells alone offered, crows did not alight.

Eastern area					S−	Es+		Es+			
Entire valley		S−	Es+	S+							S−
Western area	S+						S−		S−	S−	
Date (April)	21	21	22	22	23	23	23	24	24	25	25

Eastern area	Es+	Es+			Es−	Es−	Es+		Es+		
Entire valley										S+	
Western area			S−	S−				S−			S+
Date (April)	25	26	26	27	27	28	28	28	29	29	30

Experiment 7 We next investigated the effect of distance between egg and shell. This was done in a mass test without direct observation from the hide, and in the same valley where we had regularly seen Crows and Herring Gulls take eggs. Half-concealed gulls' eggs were laid out; one-third of them had a shell at 15 cm distance, one-third at 100 cm, and one-third at 200 cm. A total of 450 eggs were presented in 15 tests, with the results in Table 18.

Table 18. *Numbers of eggs taken and not taken with a shell 15, 100 or 200 cm away*

15 cm		100 cm		200 cm	
taken	not taken	taken	not taken	taken	not taken
63	87	48	102	32	118

No. of presentations: 3×150. Difference between 15 cm and 200 cm significant at 0·1% level. Significance of the total result: $p < 1\%$.

Part of each group of eggs may of course have been found without the aid of the shell (and this may explain why so many eggs were found even of the 200 cm group), but the figures show that the 'betrayal effect' is reduced with increased distance.

A second, similar test was taken with the shells at 15 cm and 200 cm; altogether 60 eggs were presented in 3 tests; the results are given in Table 19 and show a similar result.

Table 19. *Number of 'eggs with shell' taken and not taken for different distances (in cm) between egg and shell*

15 cm		200 cm	
taken	not taken	taken	not taken
13	17	4	26

No. of presentations: 2 × 30. Difference between 15 cm and 200 cm significant at 2·5% level.

The following test, to be mentioned only briefly because it did not contribute to our results, was in fact done before any of the tests mentioned so far in an attempt to short-circuit our procedure. We put out 50 hens' eggs painted in what we then hoped would be a good camouflaged pattern (see above), and gave each egg an egg shell dummy at 15 cm distance. These dummies were metal cylinders made by bending a strip of metal sheet measuring 2 × 10 cm as used in our later tests with Black-headed Gulls (see experiments 11, 17 and 18). Half of these were painted like the eggs, inside and outside (they were in fact satisfactorily cryptic to the human eye); the other half were painted white. We assumed that predators would be slow to come, and that in the beginning one check per day would be sufficient. Upon our first check, about 24 hr after we laid out the eggs, we found that all 50 eggs had disappeared. Our first conclusion was that a whole horde of predators, such as a flock of Herring Gulls, had raided the valley, but we later saw that this valley was searched mainly by one pair of Carrion Crows and one, two, or sometimes three pairs of Herring Gulls. During the following weeks we discovered that the Crows kept digging up these eggs which they must have buried on the first day.

This failure forced us to check the effectiveness of our 'camouflage' paint first (see experiments 1–3) and to test the effect of the natural egg shell (experiments 4–7); by the time these questions had been settled we had to start our tests on the stimuli eliciting egg shell removal in the Black-headed Gulls themselves, and so had to abandon for the moment any tests on the effect of the colour of the egg shell on the predators. However, the method having now been worked out, it is hoped to investigate this more fully.

The conclusion of this part of our study must therefore be that the eggs of the Black-headed Gulls are subject to predation; that in tests outside the colony the number of eggs found by Carrion Crows and Herring Gulls is lower than it would be if the eggs were white; that the proximity of an egg shell endangers the brood; and that this effect decreases with increasing distance. While it will now be worth investigating the predators' responsiveness to eggs and shells in more detail, the facts reported leave little room for doubt about the survival value of egg shell removal as an antipredator device. Whether or not the response has other functions is of course left undecided.

IV. The Stimuli Eliciting Egg Shell Removal

Beer's demonstration that a great variety of objects can elicit the response naturally led to the question whether all these objects did so to exactly the same extent. This was investigated by presenting dummies of different types, one at a time at each nest, to hundreds of nests, and checking after a certain period (constant for each experiment but varying for different experiments according to the overall stimulating value of the dummies compared) whether or not the dummy had been removed. Nests were marked with numbered wooden pegs, and, unless otherwise mentioned, each nest was used for only one experiment. Each experiment was arranged according to the latin square method, as set out in Fig. 100. The nests were

	a	b	c	d	e	f	g	h
A	1	8	2	7	3	6	4	5
B	2	1	3	8	4	7	5	6
C	3	2	4	1	5	8	6	7
D	4	3	5	2	6	1	7	8
E	5	4	6	3	7	2	8	1
F	6	5	7	4	8	3	1	2
G	7	6	8	5	1	4	2	3
H	8	7	1	6	2	5	3	4

Fig. 100. A latin square experiment sequence, in which 8 models (1–8) were offered in 8 successive tests (a–h) to 8 groups of nests (A–H).

divided in as many equal groups as there were dummies, and each nest was offered each dummy once; the sequence of presentation to each group was arranged in such a way that each dummy was presented equally often in the first, second, third, etc. position in the sequence.

263

Since each experiment involved repeated presentation of dummies, even though no nest had the same dummy twice in one latin square experiment, the problem of waning of the response was important. As will be discussed in detail in section v the waning within the time limits of one latin square was so slight as to be negligible, but with repetition of entire experiments with the same group of birds there was definite waning from one experiment to the next.

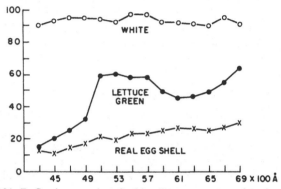

Fig. 101. Reflection graphs of white, lettuce green and real egg shell.

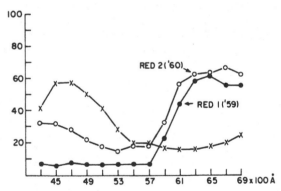

Fig. 102. Reflection graphs of blue, red 1959 and red 1960.

The dummies were put down in a uniform way on the nest's rim, namely approximately 12 cm from the centre of the nest (in 80 nests measured the average outside diameter was 27 cm, the average inside diameter 13 cm). Their orientation was likewise standardised as much as possible. Watch was kept from a hide nearby in a number of cases. From these direct observations we know that both male and female remove egg shells; we also know that egg shell dummies were not moved by the wind (quite strong gales had no effect even on some of our lightest models if they were lightly anchored on the nest's rim; yet for other reasons we avoided testing in rough weather). Small movements of the dummies might be the result of the bird's accidentally kicking it; this meant that very slight changes in the position of the dummies could not be interpreted with certainty; this category however was small and we tried to score

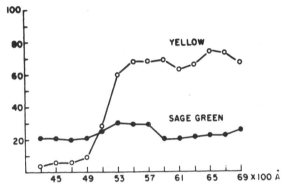

Fig. 103. Reflection graphs of yellow and sage green.

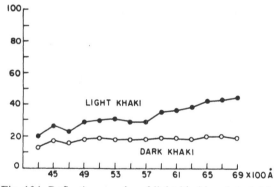

Fig. 104. Reflection graphs of light khaki and dark khaki.

them in a uniform way, displacement over more than one inch away from the nest being taken as evidence of carrying.

Because both sexes carried the egg shell, and because intervals between nest reliefs varied widely, it was impossible to keep records of individual birds, at least in the mass tests. This is an inaccuracy which, while probably not very important, has to be borne in mind in judging the results.

Since we had reasons to concentrate on the antipredator aspects of the response and particularly on camouflage, the question of visual conspicuousness, and therefore of the effect of different colours, was taken up first.

The effect of colour was tested by offering rectangles of flat sheet metal bent at right angles in the middle. In 1959 these rectangles measured 1 × 2 in.; in 1960, when we compared some of the angles with cylindrical rings of the same diameter as Black-headed Gulls' eggs, our angles were either made of strips of the same surface area as the rings, or half that size; the 'small angles' measured 2 × 5 cm; the 'large angles' 2 × 10 cm. Within each colour experiment the size of the angles was of course uniform. As colours we used Berger-master Magicote (matt) in 1959, and Rowny Fixed Powder Colours mixed with water or, for other dummies, Du-Lite emulsion paint. All paints gave a uniform matt surface. The reflected colours within the visible spectrum were measured with a Keuffel & Esser 'Colori-meter', Model E; Figs. 101–104 incl. show the reflection graphs. Any interpretation of the results obtained with such models must depend on knowledge of the sensitivity of the Black-headed Gulls' eye to light of various wavelengths. Although no data are available of this species, all species of birds examined so far are sensitive to the same spectral range as human beings; there are indications of a slightly reduced sensitivity to blue in day birds, and owls are less sensitive to red than we are, but no bird so far investigated sees infrared or ultraviolet (for a recent summary, see Schwartzkopff, **12**). Several species have been shown to distinguish well between the main colours. Weidmann (**23**), analysing the begging response of the Black-headed Gull chick, obtained scores for a variety of colours which are not compatible with colourblindness. We will assume that reflected colours which are similar to us are so to Black-headed Gulls as well, a conclusion substantiated by our own results.

Experiment 8   In our first experiment, done in 1959, the following colours were used: 'shine' (blank, shiny metal), white, red, black, 'shell' (white inside, outside sage green with dark grey dotting), sage green, 'all egg' (sage green with dark grey dotting both inside and outside), and 'cam' (sage green with dark grey irregular striping inside and outside). Shine was taken because, while its overall appearance was a rather dull grey (much darker than white) when seen from certain angles it reflected the sun and thus appeared much brighter than white. Since a bird, when moving about and when sitting on the nest, would be exposed to the bright flash far less often than to the overall greyish colour of the model, its score might help us find out when the stimulus eliciting shell removal had to be received: acting on the hypothesis (which proved correct) that white would have a high score, shine should have a lower score than white if the stimulus had to be received by the sitting bird (since the chances were then that it would usually see it as a grey angle), but if shine would have a high score, this would mean that it stimulated the bird particularly strongly whenever the bright reflection was briefly visible. White was chosen because to the human eye it was very conspicuous, and because real egg shells, particularly those which have dried for a little while, often show white patches on the rim where the pigmented shell has broken off and the egg membranes were showing (the inside of the shell is rarely white; also, most of it is in the shade). Red was chosen because there were indications in other gulls (see Tinbergen, **18**) that red objects in the nest are pecked at more often than objects of other colours. Black was originally chosen in order to check whether, if white would have a high score, this was due to whiteness or to contrast with the environment. However, it was remarkable how cryptic black really was on the nest rim; even the deep black of our dummies blended easily with the shadows of the nest rim, and as soon as one or two straws blew over a dummy or were kicked over it by the bird, a black model blended with the shades. Sage green was chosen because it was, of the available standard colours, the one most similar to the ground colour of at least some of the gulls' eggs. As is well known, the eggs of all gulls and terns show considerable variation in colour; this, in combination with the fact that the Herring Gull (and according to some of our own pilot tests, the Black-headed Gull) respond by egg rolling and by incubation to egg dummies of an extremely wide range of colours (Tinbergen, **18**, but see also Baerends, **2**) made us confident that sage green was near enough to the natural egg colour to justify its use as such—later experiments showed us how wrong we were.

This experiment was done with 96 nests, and was repeated 3 times with the same group. The results are given in Table 20.

Applying the Mann-Whitney test to the whole range of the individual trial scores (Siegel, **13**) we find the following p-values for what are clearly borderline cases: shell red: 15%; shine red: 20%; white red: 11%; shell black: 8%; white black: 4%; white all egg: 1%; cam green: 23%; black green: 3%.

We conclude that 'shine', 'white' and 'shell' elicit more responses than red and black, and that these have a higher valence than 'all egg', 'cam', and 'green'. This suggested that the contrast in brightness between the dummy and its background entirely determined its releasing value. Because, as mentioned above, there is a considerable

Table 20. *Numbers of small angles of different colours (experiment 8) carried and not carried in three successive latin square tests*[1]

	I carried	II carried	III carried	Total carried	not carried
Shine	38	34	21	93	195
White	44	25	20	89	199
Shell	38	25	18	81	207
Red	32	23	16	71	217
Black	25	24	14	63	225
All egg	27	19	12	58	230
Cam	28	13	13	54	234
Green	23	14	8	45	243

No. of presentations: $3 \times 8 \times 96$. Significance of differences discussed in text.

[1] In all tables, successive tests with the same group of birds will be indicated by roman numerals I, II, III; tests taken with different groups of birds will be indicated by capitals A, B, etc.

colour variation in the eggs of the gulls, and because previous experiments on egg recognition in Herring Gulls had shown colour to play a relatively minor part, we considered that sage green was probably similar enough to the colour of real eggs to justify the conclusion that brightness was the only colour character which controlled egg shell removal, and that the colour of the outside was not specifically responded to. The intermediate position of red and black seemed to support this, for red was physically less bright than white (Figs 101 and 102) and black was, to the human eye, remarkably cryptic. Experiments 11, 12 and 13 showed that this conclusion was incorrect. However, before discussing them, two experiments must be mentioned which were intended to examine the possible effect of contrast between the white and darker parts of the egg shell.

Experiment 9 The fact that 'shell' was not treated differently from either white or shine suggested that contrast between parts of one dummy did not play a part. In order to test this, we made a series of angles of which one 'wing' was white inside and outside, and the other wing was coloured on both sides. The additional advantage of this type of dummy compared with those of the previous experiment was that the bird would always see both colours, wherever it was; the old shell models were all white when seen from the inside, all egg when seen from the outside. We used white/white, shine/white, shell/white, sage green/white and red/white. In this experiment, 80 nests were used, and the experiment was repeated 3 times with these birds. Table 21 gives the results.

Table 21. *Numbers of small angles of different colours (experiment 9) carried and not carried in three successive latin square tests*

	I carried	II carried	III carried	Total carried	Total not carried
Shine/white	41	32	30	103	137
Red/white	34	30	22	86	154
White/white	33	30	19	82	158
Shell/white	34	26	19	79	161
Green/white	32	29	13	74	166

No. of presentations: $3 \times 5 \times 80$. Differences discussed in the text.

The Mann-Whitney test gave the following results: shine/white—shell/white: $2 \cdot 5\% < p < 5\%$; shine/white—green/white: $1\% < p < 2 \cdot 5$ shine/white—white/white $p > 5\%$; red/white—green/white $p > 5\%$.

Since in Table 20 'shine' received a higher score than 'shell', and red had a higher score than green, the results of Table 21 are consistent with the conclusion reached in the previous experiment, and there is no reason to assume that contrast within the shell-dummy plays a part. The high score of 'shell' in experiment 8 must therefore have been due to its white surface alone.

Experiment 10   Before we did the previous experiment, we had subjected 80 birds to one experiment with slightly different dummies. Two colours were used on each dummy as in experiment 9, but while colour A was on the inside of one wing, it was on the outside of the other: this ensured that both colours were always visible to the bird sitting on the nest. The combinations used were red/white, green/white, shine/white and shell/white. The scores, which we will not give in detail, are again similar except for shine/white which is a little higher than the others. Since shine had the highest score in all three experiments, it seems likely that this is a real effect; it seems most probable that this is due to its greater brightness when seen in such a way that the sun is directly reflected by it. Since this must be supposed to happen only during a very brief fraction of the time in which the bird sees the model, and in very different positions for the different birds, this again suggests that a very brief stimulus is sufficient to make the bird carry the shell away.

Experiment 11   The first indication that our interpretation had been in part incorrect came from the results of this experiment, which was taken mainly for the purpose of investigating characters of shape. We had already found (see below, experiment 16) that white angles were far less stimulating than real egg shells, and that the halved ping-pong balls which Beer had used received a score halfway between real shell and white angle. We suspected that the rounded shape of the shells had to do with this, and we therefore made cylindrical rings by bending sheet metal strips measuring 2 × 10 cm. The rings were not quite closed but had a gap of about 1 cm between the two ends. Because we had, after the 1959 season, compared the colour 'sage green' more carefully with the colours of real eggs and had seen that even the greener eggs were rather different from sage green, we had decided to use a new colour in 1960. This colour was mixed so that to the human eye it exactly matched the ground colour of the majority of eggs, which corresponds to 'ooy, 11, 4°' in Villabolos & Villabolos (21). This colour was called 'khaki'. As substitutes for real egg shells we used hens' egg shells with the 'lid' broken off. The models used in this experiment were: hens' egg shells painted khaki outside and left white inside; small rings painted khaki inside and outside, white small rings, and white angles. The results of this test, taken with two groups of nests (group A, consisting of 80 nests, was tested twice; group B, also of 80 nests, was subjected to one experiment) are given in Table 22.

Table 22. *Numbers of four models (experiment 11) carried and not carried in three tests with two groups of birds*

	A_I	A_{II}	B	Total carried	not carried
Hen's shell khaki outside	59	45	53	157	83
All white ring	36	30	44	110	130
All khaki ring	49	22	37	108	132
White angle	31	19	28	78	162

No. of presentations: 3 × 4 × 80. Difference between khaki shell and all rings: $p < 0.1\%$; difference between khaki ring and angle: $p < 1\%$.

The relevance of this test for the question of response to shape will be discussed below, but here the scores for khaki rings and white rings must be considered. They are not significantly different and in fact almost equal. If khaki represented no better 'egg colour' than sage green, there was a contradiction between this and the previous experiments. It became likely that the birds responded not only to white but also specifically to the exact colour of the eggs.

Experiment 12 We therefore decided to compare the effects of khaki and sage green in one experiment. Khaki angles were compared with green angles in 120 nests. The results are given in Table 23.

Our suspicion was strikingly confirmed: sage green received a much lower score ($p < 0.5\%$). This result raised two points: (1) how would khaki compare with the other colours? Perhaps it would turn

Table 23. *Numbers of small angles of different colours (experiment 12) carried and not carried in one test*

	carried	not carried
Khaki angles	52	68
Sage green angles	29	91

No. of presentations: 2×120.

out to be equally effective as white because it was similar to the real ground colour of the eggs; and (2) why was green so low—in experiment 8 it had received a decidedly lower score that either red or black. We thought that this was due to sage green being inconspicuous, but khaki was, if anything, less conspicuous on the nest's rim. We decided to run an entirely new and more comprehensive colour test.

Experiment 13   Because their dotting makes the overall colour of gulls' eggs darker than their ground colour, we prepared two khaki colours, one resembling the ground colour, the other, made by mixing this with some black, a darker shade of the same colour, roughly matching the overall colour of eggs seen from a distance. Unfortunately, although we had prepared a large quantity of the standard khaki colour, much of this was accidentally spilled, and we had to mix a new khaki. This was again done in such a way as to match the natural eggs' ground colour, but although we used the same ingredients our new khaki was not exactly the same as the original khaki. We do not think it likely, however, that this very slight difference has affected our results significantly but experiments using the 'new khaki' are mentioned separately. Apart from dark and light khaki our other colours were: yellow, white, a very bright yellowish green ('lettuce green'), sage green, blue, and red 2. We used small angles of 2×5 cm. The reflection graphs of these colours are given in Figs 101–104 incl. Table 24 gives the results obtained in one experiment with 240 nests; Table 25 gives the figures of an experiment with 4 of these colours, run with 80 new nests.

The Mann-Whitney test gave the following p-values: dark khaki—light khaki: 5%; white—yellow: 35%; dark khaki—yellow: $1\% < p < 5\%$; yellow—blue 5%.

Together with the results obtained in 1959 these facts point to the following conclusion: angles of any colour, however bright and unnatural, are carried to a certain extent. There are peaks for white and for khaki. Since on the nest rim (which consists mainly of dry straws) khaki is by far the most cryptic of all, and since dark khaki

Table 24. *Numbers of small angles of different colours (experiment 13) and not carried in one test*

	%	carried	not carried
Dark khaki	38	92	148
White	34·5	85	155
Light khaki	31·5	77	163
Yellow	30	73	167
Blue	25	62	178
Red	23·5	57	183
Sage green	18·5	45	195
Lettuce green	8	19	221

No. of presentations: 8 × 240.

Table 25. *Numbers of small angles of different colours (experiment 13) carried and not carried in one test*

	carried	not carried
Dark khaki	35	45
White	35	45
Yellow	28	52
Blue	22	58

No. of presentations: 4 × 80.

is carried more often than light khaki; since further the very light and conspicuous lettuce green received a very low score, the light khaki's lightness cannot be responsible for its high score, and this must be due to its special colour. In other words the birds responded to wavelength rather than brightness. Yellow has, among the bright colours with intermediate scores, the highest position, undoubtedly due to the fact that in mixing the khaki a very large amount of just this yellow had to be used; red and blue have intermediate scores, and sage green and particularly lettuce green are exceptionally low.

It thus seems that the gulls' responsiveness to colours of egg shells fits the demands very well: the two colours of the natural egg shell, khaki and white, are most stimulating, while the gulls respond least to green, the colour of the vegetation in the immediate surroundings of the nests—it would be not only wasteful, but decidedly harmful to keep carrying away the leaves of the cover round the nests.

Experiment 14   This experiment was done with small rings instead of angles. We used four colour combinations: (new) khaki inside and outside, white inside and outside, (new) khaki outside, white inside, and blank, shiny metal. The first three models were taken in order to investigate once more the possible effect of contrast within the dummy; the fourth model was added in the hope that we might find a 'supernormal' stimulus (Tinbergen, **15**). The results with 80 nests are given in Table 26.

Table 26. *Numbers of four models (experiment 14) carried and not carried in one test*

	carried	not carried
Khaki ring	37	43
White ring	39	41
Khaki/white ring	42	38
Shiny ring	40	40

No. of presentations: 4 × 80. No significant differences.

This again confirms that contrasts within the egg shell dummy have no discernible effect (experiments 8, 9 and 10). The shiny rings, while in parts brighter than white, did not give a higher score than white. New khaki and white have again approximately the same score.

Experiment 15   In this experiment, done before we began to use new khaki, old khaki, which was a little darker than new khaki, was compared, on angles, with white. Table 27, of two successive tests with 80 nests, gives a higher score for white.

Table 27. *Numbers of small angles of different colour (experiment 15) carried and not carried in two successive tests*

	I	II	Total	
	carried	carried	carried	not carried
White angle	35	36	72	88
Khaki angle	25	18	43	117

No. of presentations: 2 × 2 × 80. Difference significant at the 5% level.

In this admittedly small scale test the superiority of white over khaki, slight in experiments 13 and 14, is more pronounced. We do not know why this is so.

Experiment 16   Both in 1959 and 1960 we compared some of our models with real gulls' shells, in order to get an impression of the overall valence of the dummies used. In the experiment mentioned here, done in 1959, real gulls' shells were compared with halved ping-pong balls: still ignorant of the intricacies of responsiveness to colour discussed above, we painted them sage green with dark grey dots—roughly the same pattern as that used for 'all egg' and 'shell' angles of experiment 8. We further used white small angles of 1 × 2 in., and flat paper rectangles as used in experiment 19 reported on below. These paper strips measured 2 × 4½ cm and were a pale buff. Table 28, obtained in one experiment with 160 nests, summarises the results.

273

Table 28. *Numbers of four models (experiment 16) carried and not carried in one test*

	carried	not carried
Real shell	155	5
Halved ping-pong balls	120	40
White angle	91	69
Paper rectangle	53	107

No. of presentations: 4×160. All differences significant at the 0.1% level.

White angles therefore, while better than the buff flat strips, were clearly inferior to the halved ping-pong balls (in spite of the latter's inferior colour); and these received a lower score than real shells. This however could at least in part be due to the sage green colour of the halved ping-pong balls.

Experiment 17 Hens' shells of which the blunt end was broken off so that they resembled gulls' shells in shape while being a little larger, were painted all white, and these were compared with hens' shells painted (old) khaki outside and white inside, with real gulls' shells, and with all white small rings. This experiment differed from all others, for although each model was presented to 40 nests, each group of 40 nests received the same model on all 4 occasions (this was done to gain an impression of the degree of waning which will be discussed later). The results are in Table 29.

Table 29. *Numbers of four models (experiment 17) carried and not carried in four successive tests. No latin square arrangement; each nest received the same model throughout*

	I	II	III	IV	Total carried	not carried
White hens' shell	34	29	28	32	123	37
Khaki hens' shell	35	34	34	33	136	24
Real gulls' shell	36	35	35	33	139	21
White ring	29	28	22	28	107	53

No. of presentations: $4 \times 4 \times 40$.

There is no significant difference between real shell and khaki hen's shell; white hens' shell received a score significantly lower than real gulls' shells at the 5% level; the difference between the white ring and the white hens' shell was almost significant (p just above 5%), and that between white ring and the other shells was thoroughly significant (p $< 0.1\%$).

The conclusions to be drawn from the two last experiments are that khaki hens' shells are not inferior to real gulls' shells: that white shells are superior to white rings; and that shells are considerably superior to angles. We have already seen (experiment 11) that rings are superior to angles, and (experiment 16) that halved ping-pong balls are superior to angles. Although the colour of the paper strips was slightly different from khaki, there seems little doubt that such flat rectangles are inferior to angles.

Our next task was now to examine whether the size difference between angles and rings (which were made of metal strips twice the size of the angles) could be responsible for their different effects.

Experiment 18   We presented the following 4 models, all painted all-khaki (new): small ring (bent strips measuring 2 × 10 cm), large ring (made of a strip aluminium sheet measuring 4 × 20 cm; its weight was 16 g, that of the small ring was 10 g); the small angle made of a strip of 2 × 5 cm, and 'large angles' made of the same strips as small rings but folded at right angles in the middle. Three groups of birds were used: group A of 80 nests, group B of 80 nests, and group C of 160 nests; each group was subjected to one complete latin square. The results are in Table 30.

Table 30. *Numbers of four models (experiment 18) carried and not carried by three groups of birds*

	A	B	C	Total carried	not carried
Large ring	34	30	56	120	200
Small ring	38	40	74	152	167
Large angle	34	20	47	101	219
Small angle	31	24	52	107	212

No. of presentations: $4 \times 3 \times 320 \, (-2)$.

As expected, the small rings were carried more than small angles ($p < 0.1\%$). Small rings were carried more than large rings ($p < 2.5\%$); small rings were significantly superior to large angle ($p < 0.1\%$), but there was no significant difference between the two angles.

The relatively low score of the large rings was rather unexpected because we knew from past experience (Tinbergen, **15**, **18**; Baerends, **2**) that the egg retrieving response of gulls responds better to outsize eggs than to normal eggs. However, since we also knew that the responses to such large eggs were a combination of retrieving and avoidance (which often kept the birds away from the nest in the beginning of a test), we suspected that we might have to do with a similarly ambivalent response towards the large rings, which might, either in part or even entirely, be responsible for the difference in removal scores between large ring and small ring.

275

We therefore observed from a hide the responses of 10 individual birds to small rings and to large rings. Of these birds, 6 received the small ring first, the large second; 1 bird was presented the large ring first, the small second; and 3 birds had the small ring first, then the large ring, then the small ring once more. Their fear responses, expressed by delay in returning to the nest, alarm calls, stretching of the neck, repeated withdrawal after initial approach, were roughly classified as: no signs of fear (0), and a scale of four intensities (1, 2, 3, 4) of avoidance behaviour, 4 being the highest score. Each test lasted until the birds settled on the eggs. Table 31 summarises the results of this admittedly crude assessment.

Table 31. *Fear responses shown by ten birds presented with either a large ring or a small ring on the nest's rim. Explanation in the text*

Bird No.	Small Ring	Large Ring	Small Ring
1	0	0	
2	0	0	
3	1	1	
4	1	4	
5		3	1
6	2	3	
7	0	3	0
8	0	0	
9	0	4	0
10	0	4	0

On 9 occasions the fear score for 'large ring' was higher than for 'small ring'; it was the same (either 0 or 1) on 4 occasions, while small ring never elicited a higher score.

The lower score of these large rings is therefore in part due to the interference by another response (fear, or avoidance) with the removal response. This was the first indication of a fact which became increasingly clear as the work proceeded, viz. that we were not measuring the direct expression of the 'releasing mechanism' of just the removal response, but rather the effect of the interaction of this response with, or its dominance over, a variety of other responses. Applied to the present experiment, it might be that the large ring does stimulate the removal response as strongly as or even more strongly than the small ring. Our scores, in other words, could be said to indicate the 'resultant removal tendency'.

Experiment 19 This was done in 1959 with the special aim of investigating how the gulls distinguished between egg shells and nest material. The immediate impetus to this was given by Beer's observations, illustrated in Fig. 99, which had shown: (a) that some objects were very similar to objects which the gulls were sometimes seen to carry to their nests while building; and (b) that a bird would sometimes actually alternate between carrying an object to the nest and carrying it away. Further, some of our 'small angles' were found at the end of a test in the nest rim in such a position that it seemed as if the birds had made 'sideways building' movements with them. From a comparison of the classes of objects carried away and those found built into the nests, the main criterion of objects eliciting nest building movements seemed to be the elongate shape. This led to the use of four flat models which differed only in their proportions. Flat rectangles of thin cardboard were used of the following dimensions: 3×3 cm (model 1); $2 \times 4\frac{1}{2}$ cm (2); 1×9 cm (3); and $\frac{1}{2} \times 18$ cm (4). The surface area of all these models was the same. They were painted a buff colour rather similar to light khaki. Direct observations, and checks in which models found moved at the end of a test were classified according to the direction of the wind proved that even these light models were not blown away, not even on days with winds up to about Beaufort force 8. However, to prevent confusion through birds taking up models in their bills and dropping them, which would expose them to the wind, we decided to be on the safe side and did not run tests when the wind was stronger than about force 5.

In the first series of tests, 40 nests (group A) were offered the models in latin square arrangement 4 times in succession; in a second series, another group (B), also consisting of 40 nests, was subjected to 3 consecutive latin square tests. These tests were repeated so often because the absolute level of the score was low—this was no doubt due to the fact (see Table 28) that important characters of shape were missing in all four models. Table 32 summarises the results.

Table 32. *Numbers of four models (experiment 19) carried and not carried in seven tests with two groups of birds*

	A_I	A_{II}	A_{III}	A_{IV}	B_I	B_{II}	B_{III}	Total carried	not carried
1 (3×3 cm)	11	5	11	7	14	7	11	66	214
2 ($2 \times 4\frac{1}{2}$ cm)	8	5	10	8	15	5	7	58	222
3 (1×9 cm)	15	11	12	12	14	9	9	82	198
4 ($\frac{1}{2} \times 18$ cm)	12	9	12	7	10	10	6	66	214

No. of presentations: $4 \times 7 \times 40$.

While there is no significant difference between the scores for models 1, 2 and 4, model 3 was carried significantly more often than model 2 ($p < 2.5\%$). This was a rather unexpected result, because model 3 could not be said to approach the dimensions of the egg shell most closely; both 2 and 1 would seem to conform better.

We believe that the clue to this problem was provided by the fact that on many occasions where one of these flat models had not been removed it was found in the nest, partly covered by nest material. Table 33 shows how often a model of each category which was not carried was found in this position.

Table 33. *Numbers of models found in nest after experiment 19 compared with numbers not carried ('minus responses')*

1 (3 × 3 cm)	52 out of 214 'minus responses' 24%
2 (2 × 4½ cm)	62 228 27%
3 (1 × 9 cm)	59 198 30%
4 (½ × 18 cm)	94 206 46%

Differences between 1 and 4 and between 2 and 4 significant at the 0·1% level; that between 3 and 4 at the 0·5% level.

At first we thought that the birds had in such cases invariably treated the model as nest material and had performed sideways building movements with it. However, when we decided to check this by direct observations from a hide, the situation appeared to be even more complicated. We found that a model could land among the material of the nest either because it was taken up and dropped (which often made the impression of an incomplete carrying response), when it could subsequently be covered by the scraping and egg shifting movements of the bird; or because the bird performed actual nest building movements with it. Further, as Table 34 shows,

Table 34. *Direct observations of responses to the four models of experiment 19.*

	no notice	carried	attempts at swallowing	built-in	Total observations
1 (3 × 3 cm)	22	5	0	1	28
2 (2 × 4½ cm)	16	4	1	2	23
3 (1 × 9 cm)	10	7	2	4	23
4 (½ × 18 cm)	8	7	4	5	24

a third response was involved as well: some of the models elicited feeding. It is true that in all but one of the observations (when a model 4 was fully swallowed) the bird did not succeed in actually eating the model, but attempts at swallowing were unmistakable.

It is no doubt significant that all three responses (removing, swallowing, and building) started off with the same movement: taking the model in the bill. It was further revealing that on more than one occasion we saw a bird alternate between attempts at swallowing, incomplete building movements, and incomplete carrying movements (standing up with the model in its bill, stretching the neck, but then settling down again). What happened was therefore

278

something like this: a model might be picked up because it provided stimuli for either of the three responses. Once the bird 'found itself' with the model in its bill, this stimulus might set off either the same response as was started by the visual stimulus which elicited picking-up, or one of the two other responses. Since feeding and nest building are obviously more readily elicited by the longer models, particularly model 4, their 'carrying score' might have become higher than expected if merely egg shell removal were aroused. Model 3 might have given the highest score because it elicited more initial feeding and building responses than 2 and 3, and because it was still less exclusively eliciting feeding and nest building than the extremely long model 4. A further reason for its high score could have been that it proved more difficult to use as building material than 4, and that the response was therefore more often continued into carrying.

Whatever the exact explanation, however, the observations once more reveal the fact that the carrying score as obtained in the mass experiments reflects, not the functioning of the carrying response alone, but its preponderance over other responses, with which it interacts.

Table 35. *Numbers of hens' eggs painted khaki outside, carried and not carried when offered at different distances from nest* (experiment 20)

	carried	not carried
10 cm	64	16
15 cm	54	26
25 cm	27	53
35 cm	11	69

No. of presentations: 4×80. Difference between 10 and 15 cm not significant; that between 15 and 25 cm significant at $0 \cdot 1\%$ level; that between 25 and 35 cm significant at 1% level. Significance of the total result: $p < 1\%$.

Experiment 20 We had found (experiment 7) that the 'betrayal effect' of shells decreased rapidly with increased distance between shell and egg. It was therefore of interest to test the effect of distance on the gulls' response as well. We offered hens' shells painted khaki outside at 4 distances from the centre of the nest: on the rim (appr. 10 cm from the centre), and at 15, 25 and 35 cm distance from the centre. This was done with 80 nests: the results are given in Table 35.

It is obvious that the response decreases very rapidly with increased distance. The sharpest drop is just outside the nest's rim, between 15 and 25 cm. It is regrettable that we did not use the same

279

scale of distances in the predation tests. It looks from the predation tests as if it might 'pay' the gulls to show a better response to shells further than 15 cm from the nest, but as we shall see later there are good reasons to believe that here again another response interferes with shell removal: the need of covering the brood.

Experiment 21   When we compared the results obtained so far in analyses of stimulus situations which specifically release a single response (Tinbergen, **17**; Magnus, **8**; Weidmann, **23**) we were puzzled by the fact that, while on the one hand shell removal could be elicited by a very wide range of objects (neither shape, nor size, nor colour being strikingly specific) yet there were these two rather sharply defined peaks for khaki and white, and the very specific trough for green. We wondered to what extent this response was normally conditioned, and to what extent the selective responsiveness was independent of previous experience with eggs or egg shells. One effect of experience will be discussed below: having carried a shell slightly reduces the likelihood of carrying one again.

The present experiment was a gamble, and was done on a small scale; but the rather striking results justified it.

We knew from U. Weidmann's (**22**) work that Black-headed Gulls could be prevented from laying by offering them eggs on the empty scrape well before the first egg was due. We therefore laid out 1 black wooden egg in each of a number of scrapes in early April, some weeks before the majority of birds laid. As it turned out, some of these nests were abandoned; others were occupied by birds which did not sit on the black egg dummies but built nests on top of them; in others again eggs were laid soon after we gave the black eggs. However, 14 pairs accepted our black eggs, began to incubate on them and were thereby stopped from laying eggs of their own. When these birds began to incubate, we added 2 more black eggs to each clutch so as to ensure the best possible situation, and these 14 pairs were allowed to incubate these black clutches for approximately 5 weeks (well beyond the normal incubation period, which according to Beer (**3**) and Ytreberg (**24**), is on the average about 24 days). We then presented to these birds, in the normal latin square arrangement, black angles and khaki angles. Though we had never compared the effects of these 2 colours directly, experiment 8 had shown the black has a low valence compared with white, shine, and shell, and was about equal to red. Nevertheless we selected 14 normal pairs as controls, which had been sitting on their own eggs, and ran this experiment with them simultaneously. The results of 7 latin squares in succession, run over 5 consecutive days, are given in Table 36.

Table 36. *Numbers of small angles of different colours carried and not carried by birds which had been sitting on black eggs ('experimentals') and birds which had been sitting on normal eggs ('controls') (experiment 21)*

	Black Angle		Khaki Angle	
Experimentals	12	37	13	36
Controls	3	46	12	37

No. of presentations: $2 \times 7 \times 14$. Difference between 'experimentals' and 'controls' in their scores for 'black angles' significant at the 2·5% level.

The conclusion must be that having incubated black eggs increases the response to black egg shell dummies. We do not know the age of these birds, and hence cannot say whether they may have had experience with normal eggs in a previous season; in inexperienced birds the effect may well be still more pronounced. It is worth pointing out that these birds had had no experience with carrying black egg shells; the experience they had was with another response: incubating black eggs.

There is of course the possibility that this is not the result of a learning process at all, but that the birds at the moment of carrying matched the colour of egg shell dummy and egg: we hope to follow this question up next year.

Experiment 22   In experiment 17 we had compared real gulls' shells with hens' shells painted either khaki or white, and had been surprised to see that the scores for these three models were about equal. We thought it possible that they were not really equally effective but that even the least effective of them was still near-optimal and thus sufficient to give, under the conditions of our tests, a near 100% score. The response to shells was usually so prompt that the birds who carried at all did so immediately upon alighting on the nest. Shortening the duration of the tests was therefore hardly feasible, and we decided, after having found that the effect of a shell decreased with increasing distance, to run an experiment with several highly effective models at a greater distance from the nest. The following dummies were therefore presented at 15 cm from the centre of the nest: real gull shell; all white hens' shell; all (old) khaki hens' shells; and all lettuce green hens' shells. The results, obtained in 1 latin square with 160 nests, are set out in Table 37.

While, according to expectation, real gulls' shells were removed from more nests than any of the other models, the khaki shells were not carried more than the lettuce shells, and considerably less than white shells.[1] White was superior to khaki in our previous tests too,

[1] It should be noted however that in experiment 17 we offered hens' shells which were khaki outside but white inside whereas the present 'khaki hens' shells' were khaki on both surfaces.

Table 37. *Numbers of four different types of egg shell offered at 15 cm from nest (experiment 22) carried and not carried in one test*

	carried	not carried
Real gulls' shell	114	46
White hens' shell	80	80
Khaki hens' shell	36	124
Lettuce hens' shell	31	129

No. of presentations: 4×160. Difference between real shell and white significant ($p < 0.1\%$); difference between white and khaki significant ($p < 0.1\%$); khaki and lettuce not significantly different.

though usually not to this extent, but in the angle tests khaki had been far superior to lettuce green. The only suggestion we can offer is that with increasing distance from the nest the khaki shells, because of their strikingly cryptic colour, failed to attract the birds' attention; while a khaki model on the nest's rim cannot be missed, its inconspicuousness might have effect at greater distances, but it is certainly puzzling that this should already be noticeable at 15 cm.

V. Changes in Responsiveness in Time

As mentioned in section II, C. Beer (3) had found that Black-headed Gulls are ready to remove an egg shell throughout the incubation period and even well before the first egg is laid. Our experiments were done at different times in the breeding season, and although our conclusions are not based on inter-test comparisons, it is of interest to know as precisely as possible how the overall readiness behaves in the course of time. Dr Beer has kindly allowed us to publish the details of the tests he did to this purpose in 1958 and on which his general conclusion was based. A large number of nests were offered a halved ping-pong ball painted egg shell colour outside once every day, from 20 days before egg laying occurred till 13 days after hatching. The models were left at the nest for 6 hours, at the end of which a check was made, as in our experiments, and the models still present removed. Table 38 summarises the results.

The number of pairs tested during the first few days is naturally low since at this stage birds may still abandon their scrapes. The score for days E—20/—10 is significantly lower than that for days E—9/—1 ($p < 2.5\%$). During laying the score rises still more (p diff. E—9/—1 and E0 $< 0.5\%$). From then on it stays practically constant until the time when the chicks begin to hatch (p diff. between E18/25 and H0/6 $< 0.5\%$), when it begins to drop.

Table 38. *Summary of* C. Beer's *results with a standard model (halved ping-pong ball) offered once daily throughout the breeding season. E0: day on which first egg was laid; H0: day on which first chick hatched. Data lumped for periods of different lengths. Further explanation in text*

	E—20/—10	E—9/—1	E0	E1	E2	E3
Shell carried	174	584	97	88	91	90
Not carried	92	210	12	13	12	9
	266	794	109	101	103	99
% carried	65·4	73·7	89·0	87·1	88·3	91·0

	E4/10	E11/17	E18/25	H0/6	H7/13	H14/25
Shell carried	437	222	230	187	111	105
Not carried	47	22	29	51	80	76
	484	244	259	238	191	181
% carried	90·3	91·0	88·8	78·5	58·1	58·0

Before we can accept this as a fully reliable index of the readiness to carry at different times, we must consider the possibility that the responsiveness changes as a result of repeated presentation; it could wane, increase, stay constant, or fluctuate in a complicated way. While we have not run systematic tests with the same models throughout the season and using new birds for each test (which would have monopolised the entire colony just for this experiment alone), we have two sets of data which throw some light on this question.

First, our experiments give some information of waning under the conditions of our experiments. Table 39 gives the scores for con-

Table 39. *Scores of four successive tests in one latin square each, for fourteen latin squares*

Table	Test 1	Test 2	Test 3	Test 4	No. of Nests
9–1	43	35	36	28	96
10–1	46	39	33	27	80
X	53	35	38	39	80
11–A	46	45	43	41	80
11–B	39	43	39	41	80
13	93	62	69	68	240
14	35	28	27	30	80
15	44	36	37	41	80
17	99	114	105	101	160
26	71	66	60	64	160
18	134	126	119	126	160
19C	59	61	60	49	320
19A	39	36	29	33	320
19B	34	21	27	32	320
	835	747	722	720	2,256

secutive tests within a latin square sequence for the first 4 presentations of each of 14 latin squares, with specification of the experiments of which the total scores have been given already.

In judging these figures it should be remembered that in each of these experiments every nest received a new model on every occasion, i.e. a model that had some but not all of the characteristics of the preceding model. Whatever stimulus-specific revival of the response (Hinde, 7) there might be therefore would tend to counteract the waning of the scores. The differences between the models compared in the experiments however varied greatly: in some experiments some models scored much better than others; in others the various models used had approximately the same valence. It is therefore impossible to say more than that a certain degree of stimulus-specific change might, but must not, be reflected in these figures, and that this differs from one experiment to the other. It might be rewarding to plot waning against the degree of newness of each model as expressed through differences in absolute scores, but since we intend to continue this work we think it advisable to await more results. For our present purpose it is sufficient to point out that there is some evidence of waning, but this intra-experiment waning is at best slight.

Second, we can present some figures about waning with repreated presentation of exactly the same model, since on various occasions we repeated full latin squares one or more times with the same birds. This was done with 7 groups of birds, and the total carrying scores of each latin square are given in Table 40.

Table 40. *Positive scores in seven series of successive latin squares with the same sets of models*[1]

Experiment	I	II	III	IV	No. of presentations	Models used
8	255	177 ‖	122		2,304	angles diff. colours
9	174	147	103		1,200	angles diff. colours
11	175 ‖	116			640	shell, rings, angle
15	61	54			320	angles diff. colours
17	134	126	119	126	640	shells and ring
19A	46	25 ‖	45	34	640	cardboard rectangles
19B	53	31 ‖	33		480	cardboard rectangles

[1] A double vertical line indicates an interval of 8–10 days; all other repeats were run without interruption.

The table shows convincingly that repeated presentation of the same models does reduce the score. In some experiments this reduction is marked even when pauses of 8–10 days were given between two experiments, whereas in one of the series with (low-valence) cardboard rectangles the pause was followed by a revival. It is further perhaps significant that the figures of experiment 17 obtained with

Table 41. *Four series of scores obtained with the same models with different groups of birds at different times. Further explanation in text*

28 IV/2V	5V	2–7 V	6 V	7–10 V	10–11 V	11–13 V	14–17 V	17–18 V	19 V	21–23 V	Models
	1 61% No. = 80		2 46% No. = 80						2 46% No. = 80		khaki angles 30 min
					1 45% No. = 160		2 35% No. = 240			29% No. = 120	white angle 120 min
						2 41% No. = 80		1 57% No. = 160			white angles 240 min
2 48% No. = 96		2 48% No. = 160		1 54% No. = 160							Cardboard rectangles 240 min

high-valence objects (mainly egg shells) hardly drop at all after the first test, and show no significant fluctuation through the next three repeats.

Judging from these admittedly inadequate figures, repeated presentation of the same model must be supposed to cause some waning. This effect may have been relatively slight in Beer's work because he left the models at the nests for 6 hr—much longer than in our tests, which varied from 30 min to 4 hr—and because his standard model had a relatively high valence.

A second set of relevant data is given in Table 41.

On various occasions the same model was given at different times of the season to different groups of new birds, and the scores therefore reflect their valence at different times without interference by waning. Because the duration of exposure to the models varied from one experiment to another comparison is possible only between tests of the same exposure time, that is, between figures in the same horizontal row. Further, the tests are arranged according to calendar date. While the colony as a whole is relatively well synchronised (most birds laying their first egg between April 12 and early May, with a peak in the last week of April—C. Beer, 3), late birds do come in, and birds which have lost their first clutch do re-lay; this blurs the colony's calendar; yet on the average an arrangement of experiments according to calendar dates roughly reflects a classification according to age of eggs. In each horizontal row of Table 41 the highest score is marked 1 and the next lower score is marked 2; it will be seen that while in two rows 1 precedes 2 in time, in the two others 2 precedes 1. Whatever differences in responsiveness there are between the different groups of birds within one horizontal series, they do not seem to be correlated with age of eggs; this, as far as it goes, is in accordance with Beer's findings.

It seems most likely therefore that the readiness to remove egg shells remains roughly constant through the incubation period, with perhaps a slight tendency to increase with time, which in Beer's figures might just have been offset by whatever slight waning may have occurred. Another possibility is that waning occurs only in the first few repeats, after which the response stays at a slightly lower but further constant level. If this were so, Beer's figures would mean that the build-up before egg laying is slightly steeper than apparent, and that the lack of increase in his figures after egg laying is real.

VI. The Lack of Promptness of the Response

The predator tests reported in section III demonstrated the intense pressure exerted at least by Carrion Crows and Herring Gulls against leaving the egg shell near the nest. Admittedly our observations refer to only one colony (which however contains a sizeable part of the British breeding population), but Carrion Crows are practically omni-present and are notorious egg robbers. One should therefore expect that the Black-headed Gull would have developed a very prompt response and would remove the shell immediately after the chick has hatched. Yet, as we have seen, this is not usual. We cannot believe that the species has not been able to achieve promptness—egg shell removal is so widespread taxonomically that it must be an old response. The most likely reason is that there is a counteracting selection pressure—that too-prompt removal is in some way penalised. At first we thought that the risk of carrying the chick with the shell before it had hatched completely might be responsible, but this risk is the same for gulls and other species. Yet we have observations (admittedly few in number) which show that Oystercatchers and Ringed Plovers carry the shell with far less delay—in spite of the fact that their chicks stay in the nest for a mere couple of hours, and shell removal therefore must be less urgent. In the course of 1959 and 1960 the reason for the delay became gradually clear: in both years there were a number of Black-headed Gulls in the colony which preyed selectively on nearly hatched eggs and on wet chicks. Although we are certain that not all gulls engage in this 'cannibalism', this type of predation is very common, particularly towards the end of the season. In fact many of our efforts to observe the development of the behaviour during the first few hours of a chick's life (which were usually done late in the season) were time and again frustrated by the wet chicks' being snatched away by such robber gulls immediately after hatching. We have not made systematic notes on this, but twenty is a conservative estimate of the number of occasions on which we actually observed such chicks being taken in front of the hide, while dry chicks only hours older (equally or even more available) were left alone. The number of occasions on which we lost wet chicks without actually seeing it happen is much higher still. On only three occasions did we observe a Black-headed Gull trying to swallow a dry chick. While a wet chick is usually swallowed in a few seconds (often even in flight) dry chicks of less than a day old took approximately 10 minutes to swallow. There can be no doubt that chicks are practically safe from predation by neighbours (though not from Herring Gulls) as soon as their plumage becomes dry and fluffy.

It is interesting to see the behaviour of parent gulls sitting on hatching chicks while there are robber gulls about. The parents are aware of the latters' intentions; they show signs of increased hostility whenever the robber comes near, and they are extremely loath to leave the nest. As soon as the gulls are disturbed by other predators or fly up for human visitors robbers snatch up the wet chicks in a fraction of a second. We had the impression that the robber gulls kept an eye on many nests and knew where chicks were hatching. On one occasion we saw a still-wet chick being taken during the few seconds when the parent carried away the shell.

We feel justified therefore to ascribe the lack of promptness of the response to this tendency of some members of the colony to prey on wet chicks.

Discussion

Removal of an egg shell lasts a few seconds. It is normally done three times in a year. Nothing would seem to be more trivial than this response, which at first glance might seem to be no more than fussy 'tidying-up' by a 'house proud' bird. Yet we have seen that it has considerable survival value, and that the behavioural organisation is complicated, and well adapted to the needs. In addition, our study has given us some insight in some more general problems of an ecological and evolutionary nature. The following discussion owes much to the stimulating studies of E. Cullen (4) and J. M. Cullen (5).

Territory In 1956, N. Tinbergen (19) listed the alleged functions of territory in gulls as follows. One component, site attachment, assists in pair formation, in homing to the nest site, and in providing known shelter for the chicks. Inter-pair hostility, the other aspect of territoriality, prevents interference with pair formation, and, by forcing breeding pairs and thus nests apart, renders mass slaughter by predators less likely (see also Tinbergen, 17). We believe that the facts mentioned in this paper show a hitherto unrecognised function of territorial hostility in the Black-headed Gull: reducing the likelihood of predation by neighbouring gulls. We distinguish this effect from the effect of spacing-out on interspecific predation for the following reason. L. Tinbergen (14) has shown that some predatory birds can be guided for shorter or longer periods, by a 'searching image': they can, through an unknown process, concentrate their attention on one particular type of prey while being less responsive to other types. The stability of this state of narrowed responsiveness seems to be controlled in part by the density of the prey species: a series of successes in a short time seems to strengthen or at least to maintain specialisation, but lack of success tends to widen their responsiveness again. This property of predators, which may well be wide-spread, puts a premium on spacing-out of potential prey animals, because it increases searching time between successes. In this connection it is distance which counts.

We suggest that predation by neighbouring Black-headed Gulls is reduced by another aspect of territoriality. Each gull learns, during the prolonged period of territorial fighting in the early part of the season, not to intrude into the territories of its neighbours. The factor which reduces predation in this case is the existence of barriers, irrespective of distances between the broods (density).

Of course the inhibition of trespassing as a consequence of acquired knowledge of the boundaries does not totally prevent predation, but the fact that gulls do not trespass except very briefly and on rare moments (such as during a general alarm caused by another predator, e.g. Man or Fox which require adult gulls to leave the ground) obviously reduces the amount of intraspecies predation considerably.

The compromise character of colony density.   As we have seen, the mass attack by Black-headed Gulls discourages at least one predator, the Carrion Crow, from penetrating into the colony. This demonstrates the advantage of colonial nesting—Crows were not deterred by attacks of one, two or even three Black-headed Gulls. On the other hand, spacing-out of breeding pairs within the colony and the establishment of knowledge of boundaries, both achieved by territorial hostility, have also distinct advantages. Thus the density of a Black-headed Gull colony has the character of a compromise between at least these two opposing demands. As E. Cullen (4) and J. M. Cullen (5) have shown, a study of the anti-predator devices of other species might help in elucidating the adaptedness of interspecific differences in colony density.

Synchronisation of the breeding calendar   Some of our data strengthen the conclusion that predator pressure may be an ultimate factor in the synchronisation of breeding. Beer's data (3) show that the scatter in time of the appearance of fledged young on the beach does not differ strikingly from the scatter of egg laying. Yet there are every season a large number of late broods, partly those of birds arriving later than the main body, partly repeat clutches of birds who have lost their first clutch. Our observations show that at least intraspecific predation of wet chicks is particularly severe towards the end of the season, and it is striking how many of the late broods disappear. It is clear that most of the successful broods come from pairs which arrive early and which have not been forced to re-lay; failure to synchronise is heavily penalised.

The anti-predator system as a whole   At this stage of our studies it seems worthwhile to review what is now known of the ways in which the Black-headed Gull protects itself against predators. Some of these devices protect the individual—for brevity's sake we will call these 'egoistic' devices, even though they may at the same time protect others. Others protect the brood, often at the cost of danger to the individual, and such devices we will call 'altruistic'. Naturally these terms refer to the function, not to motivation.

The most obvious, and seemingly trivial, response is escape. Even this however takes different forms, dependent on the nature of the predator and on the age of the bird. A gull can simply fly away, as it does at the approach of a human being. The response to a Peregrine Falcon, which we have observed in detail several times, is different: the gulls fly up, form dense flocks and fly at great speed low over the ground or the water, executing quick zigzag manœuvres. We believe that 'erratic flights', in which individual gulls separate themselves from the flock and fly away, often downwards, at great speed and with very quick and sharp turns, are likely to be elicited by a Peregrine approaching gulls that are flying high. Chicks, in response to the alarmed behaviour of the adults in a colony, crouch, at first on the spot, but already after one day and occasionally even earlier they walk a little distance away from the nest and towards the surrounding vegetation. Each chick becomes soon conditioned to one or more individual hiding places; this conditioning allows them to reach safety more quickly than they would if they had to search for suitable cover.

Outside the breeding season Black-headed Gulls select wide open spaces: marshy meadows, open seashores, or water. This no doubt allows them to see an approaching predator in time.

In flocks of adult birds there operates at least one signalling system: the alarm call alerts other individuals.

The 'altruistic' habitat selection in the breeding season shows signs of anti-predator adaptedness: in the open sand dunes gulls avoid nesting on the bare sand, even though males may start by taking up a pairing territory there. Once paired however they select a nest site in moderately dense vegetation. Black-headed Gulls breeding on inland lakes usually select islands; either large ones which accommodate large numbers, or smaller islands such as individual *Molinia* bushes which offer space for one nest only. Where Black-headed Gulls nest on tidal saltings, this inclination to select islands can be the undoing of their broods in high spring tides (Tinbergen, **16**).

Several other behaviour patterns appear in the breeding season which, while endangering or at least not protecting the lives of the parents, do contribute to the safety of the brood. First, the scattering of the broods, which provides a certain degree of protection both from inter- and intraspecific predation, is effected by the balanced attack–escape system, with its components of actual attack, withdrawal, and agonistic displays (Tinbergen, **20**; Manley, **10b**). Further, unlike camouflaged species such as ducks, curlews and several other waders, the pheasants, the incubating gull leaves the camouflaged brood at the first sign of danger. The camouflage of the eggs depends on a specialised pigmentation system in the upper reaches of the oviduct.

Parent gulls attack predators; the fierceness of the attack, and the degree to which it is counteracted by escape tendencies depends on the type of predator and the resultant seems highly adaptive. From the moment the first egg is laid, at least one parent stays on the territory and guards the brood. As we have seen, egg shell removal is also effective as an anti-predator device. There can finally be little doubt that the chick's plumage protects its bearer by being camouflaged.

Thus the picture that emerges is one of great complexity and beautiful adaptedness. It has further become clear that at least some of the different means of defence are not fully compatible with each other, and that the total system has the character of a compromise between various, in part directly conflicting, demands. These conflicts are of different types. First, the safety requirements of the parents may differ from those of the brood. Thus the parent endangers itself by attacking predators. This is suggested by the fact that Foxes succeed in killing large numbers of adults in the colony. Though we have never seen a Fox killing an adult, their tracks in the sand cannot be misinterpreted. Often they kill many more birds than they eat. Some of these birds were 'egg-bound' females (Manley, **10a**), but in 1960, when we sexed 32 gulls killed by Foxes we found that 21 of these were males. Many of these gulls have their tails torn off and/or their legs broken. We believe that a Fox sometimes kills such birds by jumping at them when they 'swoop'. All this suggests that a certain balance between the tendency to attack a Fox and the tendency to flee from it is selected for.

The conflict between 'egoistic' and 'altruistic' behaviour is also very obvious in the time when the winter preference for wide open spaces changes into the preference for the breeding habitat which, as we have just seen, is dangerous to the adults. The switch towards the breeding habitat selection is not sudden; there is a long period in which the birds show that they are afraid of it; even when, after long hesitation, they settle in the colony, there are frequent 'dreads' when the birds suddenly fly off in panic; these dreads gradually subside (see also Tinbergen, **18**, and Cullen, **6**). Towards the end of the breeding season the adults begin to desert the colony in the evening to roost on the beach, leaving the chicks at the mercy of nocturnal predators.

Second, there are conflicts between two 'altruistic' modes of defence, each of which has its advantages. Crowding, advantageous because it allows social attacks which are effective against Crows, has to compromise with spacing-out which also benefits the broods.

Finally there may be conflicts between the optimal ways of dealing with different predators. Herring Gulls and Crows might be prevented entirely from taking eggs and chicks if the gulls stayed on the nests, but this would expose them to the Foxes. While Herring Gulls and Crows exert pressure towards quick egg shell removal, neighbouring gulls exert an opposite pressure; the timing of the response is a compromise.

We cannot claim to have done more than demonstrate that egg shell disposal is a component of a larger system, nor are we forgetting that much in our functional interpretation requires further confirmation. It seems likely however that a more detailed study of all the elements of anti-predator systems of this and other species, and of the ways they are functionally interrelated, would throw light on the manifold ways in which natural selection has contributed towards inter-specific diversity.

VIII. Summary

The Black-headed Gull removes the empty egg shell shortly after the chick has hatched. The present paper describes some experiments on the function of this response, and on the stimuli eliciting it. Carrion Crows and Herring Gulls find white eggs more readily than normal gulls' eggs; it is concluded that the natural colours of the eggs afford a certain degree of cryptic protection. When normal eggs are given an egg shell at 15 cm distance their vulnerability is greatly increased; this 'betrayal effect' decreases rapidly with increased distance between egg and shell. We therefore conclude that egg shell removal helps to protect the brood from predators.

As reported by C. Beer (**3**), the Black-headed Gull removes a surprisingly wide range of objects from the nest. Large scale tests with egg shell dummies in which colour, shape, size and distance from the nest were varied showed that objects of all colours are carried but that 'khaki' (the normal ground colour of the egg) and white are particularly stimulating, while green elicits very few responses. Egg shells elicit more responses than cylindrical rings of the same colour, and these are responded to better than 'angles'. Size can be varied within wide limits; very large rings elicit fear which interferes with removal. Various other indications are mentioned which show that the score as obtained in the mass tests does not accurately reflect the responsiveness of the reaction itself but rather the result of its interaction with other behaviour tendencies. The eliciting effect decreases rapidly with increasing distance.

On the whole, the gulls' response is very well adapted to its main function of selectively removing the empty shell, but the relatively high scores for objects which have very little resemblance to egg shells suggest that it is adapted to the removal of any object which might make the brood more conspicuous.

A pilot tests showed that gulls which have incubated black eggs respond better to black egg shell dummies than normal gulls.

The lack of promptness of the response as compared with non-colonial waders (Ringed Plover and Oystercatcher) is adaptive, since it tends to reduce predation by other Black-headed Gulls, which are shown to prey selectively on wet chicks. A hitherto unrecognised function of territory is suggested.

In a discussion of the entire anti-predator system of the Black-headed Gull its complexity and its compromise character are stressed: the safety demands of the individual clash with those of the brood; there are conflicts between the several safety devices which each benefit the brood; and there are clashes between the ideal safety measures required by each type of predator.

REFERENCES

1 BAERENDS, G. P. (1957). 'The ethological concept "Releasing Mechanism" illustrated by a study of the stimuli eliciting egg-retrieving in the Herring Gull', *Anatom. Rec.*, **128**, 518–19.

2 —— (1959). 'The ethological analysis of incubation behaviour', *Ibis*, **101**, 357–68.

3 BEER, C. (1960). 'Incubation and nest-building by the Black-headed Gull', D.phil. thesis, Oxford; subsequently published; see BEER, C. (1961), *Behaviour*, **18**, 62–106; (1963) *Behaviour*, **21**, 13–77; (1963) *Behaviour*, **26**, 189–214.

4 CULLEN, E. (1957). 'Adaptations in the Kittiwake to cliff-nesting', *Ibis*, **99**, 275–302.

5 CULLEN, J. M. (1960). 'Some adaptations in the nesting behaviour of terns', *Proc. 12th Internat. Ornithol. Congr. Helsinki 1958.*

6 —— (1956). 'A study of the behaviour of the Arctic Tern, *Sterna paradisca*', D.phil. thesis, Oxford.

7 HINDE, R. A. (1954). 'Factors governing the changes in strength of a partially inborn response, as shown by the mobbing behaviour of the Chaffinch (*Fringilla coelebs*): II. The Waning of the response', *Proc. Royal Soc.*, B **142**, 331–58.

8 MAGNUS, D. (1958). 'Experimentelle Untersuchungen zur Bionomie und Ethologie des Kaisermantels *Argynnis paphia* L. (Lep. Nymph.). I', *Zs. Tierpsychol.*, **15**, 397–426.

9 MAKKINK, G. F. (1936). 'An attempt at an ethogram of the European Avocet (*Recurvirostra avosetta* L.)', *Ardea*, **25**, 1–62.

10a MANLEY, G. H. (1957). 'Unconscious Black-headed Gulls', *Bird Study*, **4**, 171–2.

10b —— (1960). 'The agonistic behaviour of the Black-headed Gull', D.phil. thesis, Oxford.

11 NETHERSOLE THOMPSON, C. and D. NETHERSOLE THOMPSON (1942). 'Egg-shell disposal by birds', *British Birds*, **35**, 162–69, 190–200, 214–24, 241–50.

12 SCHWARTZKOPF, J. (1960). 'Physiologie der höheren Sinne bei Säugern und Vögeln', *Journ. f. Ornithol.*, **101**, 61–92.

13 SIEGEL, S. (1956). *Nonparametric Statistics for the Behavioral Sciences*, New York.

14 TINBERGEN, L. (1960). 'The natural control of insects in pine woods: I. Factors influencing the intensity of predation by songbirds', *Arch. néerl. Zool.*, **13**, 265–336.

15 TINBERGEN, N. (1951). *The Study of Instinct*, Oxford.

16 —— (1952). 'When instinct fails', *Country Life*, Feb. 15, 412–414.

17 —— (1952). 'On the significance of territory in the Herring Gull', *Ibis*, **94**, 158–59.

18 —— (1953). *The Herring Gull's World*, London.

19 —— (1956). 'On the functions of territory in gulls', *Ibid*, **98**, 408–11.

20 —— (1959). 'Comparative studies of the behaviour of gulls (Laridae); a progress Report', *Behaviour*, **15**, 1–70.

21 VILLABOLOS, C. and J. VILLABOLOS (1947). *Colour Atlas*, Buenos Aires.

22 WEIDMANN, U. (1956). 'Observations and experiments on egg-laying in the Black-headed Gull (*Larus ridibundus* L.)', *Brit. Journ. Anim. Behav.*, **4**, 150–62.

23 WEIDMANN, R. and U. WEIDMANN (1958). 'An analysis of the stimulus situation releasing food-begging in the black-headed gull', *Anim. Behav.*, **6**, 114.

24 YTREBERG, N. J. (1956). 'Contribution to the breeding biology of the Black-headed Gull (*Larus ridibundus* L.) in Norway', *Nytt Magas. Zool.*, **4**, 5–106.

(From the Department of Zoology, Oxford University)

7

Egg Shell Removal by the Black-headed Gull (*Larus ridibundis* L.):

The Effects of Experience on the Response to Colour[1] (1967)

Introduction

Black-headed Gulls remove the empty egg shell shortly after the hatching of the chick. An investigation into the survival value of this response and of the stimuli eliciting it (Tinbergen *et al.*, **4**) yielded the following results. An empty egg shell at a distance of 10 cm from a single egg laid out in the dunes renders such an egg more vulnerable to predation by Carrion Crows (*Corvus corone*) and Herring Gulls (*Larus argentatus*); this suggests that egg shell removal may help to reduce predation.

The response can be elicited throughout the incubation period by a variety of objects in the nest or on its rim, in fact by 'any object which does not resemble an egg, a chick, or nest material' (Tinbergen *et al.*, **4**) although the egg shell itself is optimal. In systematically conducted tests with dummies, it was found that colour, shape, size and distance between nest and shell affected the response.

These results suggested, therefore, that this seemingly trivial response, which normally takes no more than 20 sec each year of a bird's time, contributes to the survival of the brood; and they showed that it is controlled by a complicated and well-adapted mechanism.

A pilot test, in which the response to black and khaki models was tested in birds which had been prevented from laying any eggs that season and had been sitting instead on black eggs, showed that such gulls responded more to black models than control gulls who had

[1] Our thanks are due to Sir William Pennington Ramsden, Bt, and the Cumberland County Council for permission to work in the Ravenglass gullery; and to the Nuffield Foundation and the Nature Conservancy for support. M. Paillette wishes to express her gratitude for support received from the British Council and the Zoological Society of London.

been incubating their own, khaki coloured eggs. It was decided to investigate this with larger numbers of gulls and with green eggs as well, and also to examine whether this increased response was due to a learning process or to the birds carrying objects of a colour which at the moment of the test matched the colour of its eggs. In addition, we decided to use these same birds to investigate the effect of this type of experience on egg rolling, or retrieving.

Methods

On April 18, 1961, when very few gulls had started to lay, we selected 120 empty 'scrapes'. In 60 of these we put a plaster-of-paris dummy of a Black-headed Gull's egg painted a matt black; the other 60 received a similar egg dummy painted 'sage green' (for reflection graphs of the colours used in egg and egg shell dummies, as well as of the colour of the natural eggs, refer to Figs 101, 103 and 104 of the preceding paper). A number of gulls accepted these dummy eggs and incubated them; these received a second egg dummy on the next day, and a third dummy on the third day. Other pairs deserted the nest as a consequence of our interference; the dummies from such nests were taken away and deposisted in another empty scrape. Since such desertions occurred at a number of scrapes, the days on which incubation finally started on the individual nests varied considerably. On May 9th we stopped selecting new scrapes; between then and May 12th when we started testing, some nests were deserted and not replaced; hence the final numbers of nests tested in each group were slightly below 60, namely, 56 nests with black eggs and 53 nests with green eggs. Some of these pairs were inhibited from laying by accepting our dummy eggs (see Weidmann, 5) and thus had no naturally coloured eggs of their own in 1961; this was so with 15 pairs which had received black eggs and with 20 pairs which had accepted green eggs. In all, scrapes which received black eggs were deserted in 43 cases, those which had received green eggs were deserted in 32 cases. Hence, it seems that green eggs were more readily acceptable than black eggs. The 41 ($= 56 - 15$) pairs with black eggs which were not prevented from laying were always robbed by us of any natural egg that appeared in their nests; the same was done with the 33 ($= 53 - 20$) pairs which started laying after having received a green egg. Our two groups were thus partly composed of birds which had not laid eggs in 1961, partly of birds which had laid but had only had brief experience with their own eggs. The mean time during which the pairs had been incubating when our tests began was 16·2 days for the 'black egg birds' and 17·6 days for the 'green egg birds' (the normal incubation time is approximately 24 days (Beer, **2**; Ytreberg, **6**).

Our third group, consisting of control birds which had been allowed to keep their own eggs, had not been checked as to their time of laying; however, since the peak of egg laying occurred at about April 25th, they must be assumed to have incubated on the average during approximately the same time as the black egg and green egg groups.

In the experiments each nest received three differently coloured egg shell dummies, one at a time. Six tests were done at each nest, so that each model was offered twice. As egg shell dummies we used strips of thin metal measuring 2×10 cm and bent in a flat cylinder ('small rings' of our previous experiments); these cylinders had about the same diameter as real eggs. The three colours used were khaki (which resembles very closely the ground colour of real eggs), the same black paint as used for the black egg dummies, and the same green colour as used for the green egg dummies. The six tests were arranged as given in Table 42. Each model was left on the nest's rim for one hour, after which models were scored as either 'carried away' or 'not carried'.

Table 42. *Test schedule for the three sub-groups, A, B, and C, of each of sixty nests*

	Test 1	2	3	4	5	6
A	green	khaki	black	khaki	green	black
B	black	green	khaki	green	black	khaki
C	khaki	black	green	black	khaki	green

This method was not ideal. The three main flaws were these: (1) the birds had to be disturbed for each test, and this may in some cases have caused them to desert the clutch. That this actually happened was indicated by the fact that even towards the end of our test period freshly laid eggs appeared in some of the nests, which must have been laid by new birds which had replaced the original owners. (2) At the end of each test several of our rings were found in the nest. This, as we know from direct observations from hides, may occur in different ways. A bird may take up a model without carrying, and drop it again on the spot when it may roll into the nest; this is probably a low-intensity carrying response. A bird might also roll a ring into the nest as if it were an egg. Finally, a ring (or an egg) may be accidentally kicked into the nest when the bird shifts its clutch or scrapes while settling, or when it flies off the nest in panic. We had no means of deciding between these possibilities and therefore scored as 'carried' only those models which had been actually removed. Eggs were scored as 'rolled in' when they had moved into the nest. There is no reason to suppose that the three groups differed substantially in the incidence of such doubtful responses. We would no doubt have obtained purer results if we had observed every single test from hides; this, however, would have been so time-consuming that it would have been totally impossible to obtain scores fit for statistical treatment. (3) Both sexes incubate, and in these mass tests we had no means of deciding which partner had been subjected to our test situation. This again was cancelled by the large number of nests in each group.

Results

Table 43 summarises the results of this first experiment. The significance of the differences was tested with the Mann-Whitney U-test, as described in Siegel (3).

Table 43. *Numbers of khaki, black, and green rings removed by the three groups*

Group of nest	Model presented	Carried	Number	Percentage carried
Real-egg group (*Rr*)	khaki	71	120	59·0
	black	50	120	41·5
	green	34	120	28·5
Black-egg group (*Bb*)	khaki	42	112	37·5
	black	51	112	45·5
	green	42	112	37·5
Green-egg group (*Gg*)	khaki	41	106	38·5
	black	40	106	38·0
	green	52	106	49·0

The differences in the scores for khaki and black between the real egg group (*Rr*)[1] and the black egg group (*Bb*) were significant at the 1% level; those in the scores for khaki and sage green between the real egg group (*Rr*) and the green egg group (*Gg*) were significant at the 0·1% level; and those in the scores for black and sage green between the black egg group (*Bb*) and the green egg group (*Gg*) were significant at the 2% level.

When tested in this way, therefore, each group carried more shell dummies of their own eggs' colour than the other groups did, and even more of their own colour than of the other colours. This result could have been effected in either of three ways: (1) the birds could have been conditioned to their 'own' egg colour and had transferred

[1] In the symbols characterising the birds at the moment of the test ('*Rr*', '*Bb*', '*Gg*', etc.) the first (capital) letter refers to the colour of the eggs they had been sitting on before they were tested; the second to the colour of the eggs they had in their nests during the test.

this to the shell-removal response; or (2) the birds could, at the moment of testing, have matched the shell dummy's colour against that of the eggs present in the nest at that moment, and removed preponderantly matching shell dummies; or (3) both effects were operating. In order to test this, we repeated the tests, but with this difference, that the birds were given, during the test only, the eggs of a different colour from those they had been incubating. For practical reasons (exchanging eggs, in addition to depositing shell dummies, disturbed the birds for a much longer time than in the first experiment) we did this experiment with the black egg group and the green egg group only. The results are set out in Table 44.

Table 44. *Numbers of black and green rings removed by the black-egg birds and the green-egg birds when sitting on the 'wrong' colour*

Group of nests	Model presented	Carried	Number	Percentage carried
Black-egg group, green eggs during test (*Bg*)	black	42	110	38·0
	green	37	111	33·5
Green-egg group, black eggs during test (*Gb*)	black	29	95	30·5
	green	38	95	40·0

The differences in scores for black and green between the two groups in this experiment were significant at the 2% level; those in scores black-green between '*Bg*' and '*Gg*' (Table 44) are not significant (p = 8·6%); those in scores black and green between '*Gb*' and '*Bb*' are not significant either (p > 8%).

Therefore, having the 'wrong' colour of eggs in the nest during the test did not significantly alter the colour preference, and the birds kept removing predominantly the models which matched the colour of the eggs they had been sitting on before the test. We must conclude that our first hypothesis is correct, i.e. that having incubated on khaki, black or green eggs results in a change of the birds' responsiveness to the colour of the egg shells when they remove them; in other words, that experience gained during incubation is transferred to another response. As in normal birds, the response is not strictly confined to one colour, but neither is it confined to a particular shape, nor size (Tinbergen *et al.*, **4**).

Effect of Experience on Egg Retrieving We used the same three groups of birds to test whether experience with abnormally coloured eggs would also influence the birds' egg retrieving response. Each nest was given, immediately after each egg shell removal test, a natural, black or sage green egg on the nest's rim. The distance between egg and centre of the nest was 15 cm and, when necessary, a sand platform was built so that the egg would have to be rolled over a horizontal surface. This egg had the same colour as the shell dummy which had been presented immediately before. After 30 minutes all nests were inspected and those where the egg was found in the nest were marked 'rolled in'. The results are given in Table 45.

Table 45. *Numbers of natural, black and green eggs rolled in by the three groups*

Group of nests	Egg model presented	Rolled in	Number	Percentage rolled in
Real-egg group (*Rr*)	natural	89	120	74·0
	black	48	120	40·0
	green	39	120	32·5
Black-egg group (*Bb*)	natural	80	112	71·5
	black	61	112	54·5
	green	51	112	45·5
Green-egg group (*Gg*)	natural	74	106	70·0
	black	31	106	29·0
	green	66	106	62·5

Again, there is a clear shift towards the colour of the eggs on which the birds had been sitting, but this shift is far less pronounced than in the shell-removal responses. The differences in the scores for natural and green between *Rr* and *Gg* are significant at the 1% level; those in the scores for black and green between *Bb* and *Gg* are significant at the 2% level, and those in the scores for natural and black between *Rr* and *Bb* at the 4% level.

It is striking that, while in spite of the shorter test time, the scores for rolling and removal are of the same order as far as the black and green models are concerned, the retrieving score for khaki is much higher than the removal score for the same colour; in the retrieving tests 'natural' was preferred even by the 'black egg birds' and the 'green egg birds'. The most likely explanation would seem to be that the real eggs provide some stimuli (such as dotting and shell texture) which were not provided by the black and green eggs. While dotting does affect egg retrieving, at least in the Herring Gull (Baerends, **1**) there are no indications that it affects egg shell removal in the Black-headed Gull (Tinbergen *et al.*, **4**).

Next we investigated whether the shift in colour preference is really the result of experience. As in the shell removal tests, the black egg birds and the green egg birds were tested with black and green eggs while they were, for the duration of the tests, sitting on eggs of the 'wrong' colour. These tests gave the results presented in Table 46.

The differences in black-green scores between the two groups in this experiment are not significant (p = ·56%), but neither are those be-

Table 46. *Numbers of black and green eggs rolled in by the black-egg birds and the green-egg birds, while sitting on the 'wrong' colour*

Group of nests	Egg model presented	Rolled in	Number	Percentage rolled in
Black-egg group	black	79	107	74·0
green eggs in test (*Bg*)	green	76	112	68·0
Green-egg group,	black	67	95	70·5
black eggs in test (*Gb*)	green	72	96	75·0

tween *Gb* (Table 46) and *Bb* (Table 45) (p = ·24%). However, *Bg* (Table 46) has significantly different scores from *Gg* (p = 5%). This means that the scores for black and green, which were different in Table 45, rather levelled out (though not completely) when the birds were put on the 'wrong' eggs. This might be due to a 'matching effect' in the last experiment, but, since these experiments involved 1½ hours' exposure to the 'wrong' eggs every time the birds were tested for both shell removal and egg retrieving (6 times in all, i.e. a total of 9 hours' exposure) it is also possible and, in fact, rather more likely that the conditioning to black and green eggs was gradually broken down. That this conditioning was real is indicated by the fact that *Bg* responded very differently from *Gg*. Why the effect is not equal in both groups cannot be said with certainty, but it would seem to be significant that in the tests of Table 45 the 'green egg group' had a high score for green eggs, while the 'black egg birds' rolled in hardly more black eggs than green eggs. The entire effect might well be due to a high valence of green eggs (which Barends, 1, demonstrated in the Herring Gull). As pointed out by Tinbergen *et al.* (4) the low valence of green in *shell removal* must have survival value because it prevents the birds from removing leaves of the surrounding cover. No such pressure against responding to green eggs by retrieving can be expected.

Our results therefore show that a few weeks' incubation on abnormally coloured eggs influences the colour preferences shown in both egg retrieving and egg shell carrying. It is peculiar that this effect should be as clear in egg shell removal as in egg retrieving, because it is sitting on and shifting of the eggs which are directly involved in the birds' experience with the abnormal eggs, not shell removal—the response to which the conditioning is transferred.

We do not believe that these results can be correctly described by saying that the response to colour in egg shell removal is normally conditioned. Our experiments with a wider range of colours presented to normal birds (see Tinbergen *et al.*, 4) do not merely show a high score for khaki (the colour of the eggs on which such birds had been incubating) but also for white—a colour of eggs the birds have not

MODELS → IN NEST	KHAKI	BLACK	GREEN	NAT L.	BLACK	GREEN
NAT L.	59%	41.5	28.5	<u>74</u>	40	32.5
BLACK	37.5	³⁸ <u>45.5</u>	^{33.5} 37.5	71.5	⁷⁴ <u>54.5</u>	⁶⁸ 45.5
GREEN	38.5	^{30.5} 38	⁴⁰ <u>49</u>	70	^{70.5} 29	⁷⁵ <u>62.5</u>

Fig. 105. A graphic summary of the results, giving the percentages mentioned in the four tables. The highest figures in each column are underlined. The smaller figures in the upper right corners of eight compartments refer to the scores of Tables 44 and 46 when the nests contained eggs of the 'wrong' colour.

had during incubation; moreover, such birds have an extremely low score for green. For a complete evaluation of the part played by experience, it is further of great interest that three individual birds which were given an egg and, on a separate occasion, an egg shell on the nest rim before they had had any experience with eggs or shells (i.e. birds in immature plumage which were tested when their first scrape was still empty) showed both removal response and egg retrieving quite normally, although colour preferences have not been tested in these birds. We hope to be able in the future to test birds on a much larger scale.

The only conclusion which seems warranted at the moment is that colour preference in egg retrieving and in egg shell removal can be influenced by forcing the gulls to incubate abnormally coloured eggs; this experience increases their responses to the colour of the 'adopted eggs'.

Summary

1. An attempt was made to influence the valence of different colours for the elicitation of egg shell removal (see Tinbergen et al., **4**) by forcing the birds to incubate abnormally coloured eggs. In addition, the effects of this treatment on colour preference in egg retrieving were studied.

2. One hundred and eighty nests were divided in three groups of which one retained the eggs laid by the birds themselves, another received black egg dummies, and a third was given green egg dummies. After about seventeen days' incubation the responses to egg shell dummies coloured 'khaki', black, and green were recorded in these three groups, as well as the retrieving of black, green and real eggs.

3. In both test series the gulls showed an increased response to dummies of the colour of their 'adopted' eggs.

4. This effect was retained when the birds were given, during the tests only, eggs of a different colour from those they had been incubating. This showed that the birds' responses were affected by experience rather than by similarity between the dummies and the clutch in the nest at the moment of the test (Fig. 105).

5. Three immature gulls which were given egg shells and eggs on the rim of their scrapes before they had laid eggs showed normal removal and retrieving responses to the appropriate object. This shows that the response is not entirely acquired.

6. While the experiments show that the effect of colour on these two responses can be influenced by experience during incubation, neither the high score for white nor the low score for green in the removal of normal birds can be explained by this experience.

REFERENCES

1 BAERENDS, G. P. (1957). 'The ethological concept "Releasing Mechanism" illustrated by a study of the stimuli eliciting egg-retrieving in the Herring Gull', *Anatom. Record*, **128**, 518–19.

2 BEER, C. G. (1961). 'Incubation and nest building behaviour of Black-headed Gulls: I. Incubation behaviour in the incubation period', *Behaviour*, **18**, 62–107.

3 SIEGEL, S. (1956). *Non-parametric Statistics for the Behavioral Sciences*, New York.

4 TINBERGEN, N., G. J. BROEKHUYSEN, F. FEEKES, J. C. W. HOUGHTON, H. KRUUK and E. SZULC (1962). 'Egg shell removal by the Black-headed Gull, *Larus r. ridibundus* L.; a behaviour component of camouflage', *Behaviour*, **19**, 74–117.

5 WEIDMANN, U. (1956). 'Observations and experiments on egg-laying in the Black-headed Gull (*Larus ridibundus* L.)', *Brit. J. Animal Behav.*, **4**, 150–61.

6 YTREBERG, N. (1956). 'Contribution to the breeding biology of the Black-headed Gull (*Larus ridibundus*) in Norway', *Nytt Mag. f. Zoologi*, **4**, 5–106.

(From the Department of Zoology, Oxford University)

8
How do Black-headed Gulls Distinguish Between Eggs and Egg Shells?[1]
(1962)

Introduction

Black-headed Gulls (*Larus ridibundus*) remove the empty egg shell shortly after the hatching of the chick. An investigation into the survival value of this response and of the stimuli eliciting it (Tinbergen *et al.*, **2**) yielded the following results. An empty egg shell at a distance of 4 in. from a single egg laid out in the dunes rendered such an egg more vulnerable to predation by Carrion Crows (*Corvus corone*) and Herring Gulls (*L. argentatus*); this suggests that egg shell removal may help to reduce predation. The response can be elicited throughout the incubation period by a variety of objects in the nest or on its rim, in fact by 'any object which does not resemble an egg, a chick, or nest material', although the egg shell itself is optimal. In systematically conducted tests with dummies, it was found that colour, shape, size and distance between nest and shell affected the response.

These results suggested, therefore, that this seemingly trivial response, which normally takes no more than 20 sec of a bird's time each year, contributes to the survival of the brood; and they showed that it is controlled by a complicated and well-adapted mechanism.

This first study, while not allowing us to list all the characteristics of the situation which elicits egg shell removal, showed that the birds distinguish between egg shell and nest material by at least four

[1] Our thanks are due to Sir William Pennington Ramsden, Bt, and the Cumberland County Council for permission to work in the Ravenglass gullery; and to the Nuffield Foundation and the Nature Conservancy for support. M. Paillette wishes to express her gratitude for financial assistance received from the British Council and the Zoological Society of London; R. A. Stamm was helped by a grant from the Janggen-Pöhn Stiftung.

characteristics: in fact, the egg shell is distinguished by being three-dimensional, rounded, less oblong than nest material, and partly white.

The present paper deals with the question of how gulls distinguish between egg shells and eggs. Failure to do so would naturally endanger the brood, and we have actually never observed a gull removing its eggs or chicks.

Although an intact egg and an egg shell as left after the chick has hatched have much in common, they elicit two entirely different sets of responses. An egg in the nest is brooded and occasionally shifted. If an egg happens to lie on the nest's rim or even a little outside the nest, the bird often retrieves it. It may either sit down first or remain standing in the nest, then may gently touch the egg's upper or distal surface with the ventral side of its bill. This is then often followed by 'rolling in', which is done by bending over the egg and balancing it against the under surface of the lower mandible, then rolling it back towards the breast, i.e. into the nest cup (Fig. 98).

A shell in or near the nest is taken in the bill (Fig. 98). The bird gets hold of the thin edge and either walks or flies away with it at once, or nibbles it for a while before removing it. Nibbling may be interrupted when the bird drops the shell, but it ultimately leads to 'carrying'. Sometimes a gull pecks at the material which is left behind in the shell and it may eat some of it.

The situations eliciting the two responses in our experiments were the same except for the objects (egg and egg shell) themselves.

The differences we see between an egg and an egg shell could be described in various ways, and the following descriptions guided us in our attempts to find out to which properties the gulls reacted:

Egg	Egg shell
(1) Smooth, oval outline	Interrupted, partly serrated outline
(2) Closed; no opening	With opening; hollow
(3) No thin edge or rim	A thin edge
(4) Heavy (appr. 37 g)	Light (appr. 2 g)
(5) Dotted khaki; no white	White on rim and inside
(6) Total egg-coloured surface	Small egg-coloured surface

Of these, all except (4) could be perceived visually. Since both responses are initiated by visual stimuli (the birds focusing the objects from a distance, irrespective of the direction of the wind, and visually similar dummies being effective), weight could not play a part in the release of the response, but it might have an effect once the bird has made contact.

Method

This study was carried out at the Black-headed Gull colony at Raven-glass, Cumberland, in the summer of 1961. Various models, to be described later, were presented, one at a time, on the rims of nests containing two or three eggs, and the birds' responses were observed from hides approximately 8 ft away. Each bird was given each model only once; and each was tested with all models of a series. The sequence of presentation was varied at the different nests, as much as possible according to a latin square arrangement. Each individual test lasted either 15 mins or—if a bird completely performed one of the two end-acts (rolling-in or carrying) before the 15 min were over—until either end-act had been completed. After each test, the observer signalled to an outside helper without disturbing the bird. The helper then approached, inevitably forcing the bird to fly up, and as quickly as possible prepared the next test situation. With each nest the models compared were always put out in the same place.

We could usually distinguish the two partners of a pair and were able, therefore, to test individual birds. However, because we could not control the moment of nest-relief, some series had to be broken off before the relieved bird had been tested with the entire series of models. The number of presentations varies slightly in each experiment, therefore.

The following terms were used to characterise the birds' behaviour:

Billing: touching the model's surface with the bill, which is usually closed (this generally precedes rolling).

Intention-rolling: bending the neck over the model and touching its distal surface without actually moving it.

Rolling: pushing the model in the direction of the nest cup, irrespective of success.

Lifting: picking up the model, then dropping it without carrying.

Nibbling: mandibulating the rim of the model.

Intention-carrying: getting hold of the model with a stronger grip, and often dragging it a little way.

Carrying: picking up the model, walking or flying away and dropping it at some distance from the nest.

Poking: pecking into the hollow model, sometimes eating material left inside.

Ignoring: making no response to the model, apart from looking at it.

(For the statistical tests employed, which are mentioned separately below, see Siegel, **1**.)

Experiment 1   This was a pilot experiment in which no more than the two end-acts (carrying and rolling) were recorded; it was designed to see whether clear-cut results were at all possible with small numbers of birds. Empty eggs with holes of different sizes were used in this test. For reasons not to be discussed here, the holes were made at the sharp ends of the eggs (the hatching chick invariably cuts off a 'lid' at the blunt end) and the edge of each hole was smooth, not serrated. We used eggs of approximately the same size; the size of each hole was expressed indirectly in the projected distance between its edge and the blunt pole of the egg.

The following models were used: 'Small hole' (empty egg with edge of hole 51 mm from end); 'Medium hole' (similar but 45 mm); 'Large hole' (similar again but 36 mm); and 'Real shell' (egg shell from which a chick had actually hatched the previous year). The diameter of 'Large hole' was approximately the same as that of the hole of a real shell. Each model was presented to 12 birds. Table 47 summarises the results.

Table 47. *Results of experiment 1*

	Rolling	Carrying	Neither	Total presentations
(b) Small hole	9	1	2	12
(c) Medium hole	9	1	2	12
(d) Large hole	4	6	2	12
(h) Real shell	0	9	3	12

For computing the significance of these results, the Fisher exact-probability test was used. The scores for 'Small hole' and 'Medium hole' are significantly different from those for 'Large hole' ($p < 5\%$). Thus we reached three conclusions. First, a real shell is not rolled in (though in other tests it sometimes was, as can be seen from, for example, Table 48); when the bird responds to it at all, it removes it. Second, the hole must be fairly large to make the bird switch over to carrying in part of the tests. Third, a real shell elicits more carrying and less rolling than even 'Large hole'.

These results were, however, difficult to interpret. We did not know whether the birds reacted to the size of the hole, the hollowness, the extent of the rim, the surface area of khaki colour or what; nor did we know whether the difference between 'Real shell' and 'Large hole' was due to the serrated edge or to the former's showing more white.

Experiment 2   We next presented the following four models to new

Fig. 106. From left to right: normal egg shell; egg shell filled with plaster of paris; egg shell with cotton wool; egg shell with piece of lead; 'flanged' egg.

birds, and observed their responses in more detail: 'Empty egg' (a blown-out egg, with hardly visible holes at both ends); 'Filled shell' (a real egg shell filled to the rim with plaster of paris, so that it was not hollow and did not show a thin edge, but did have an interrupted and partly serrated contour); 'Shell-with-lead' (a real egg shell with, inside and at the obtuse end, a piece of white-painted lead equal in weight to a real egg); and 'Real shell' (as in experiment 1). 'Shell-with-lead' was heavier than 'Filled shell'. These models are illustrated in Fig. 106. Table 48 summarises the results.

Table 48. *Results of experiment 2*

The responses of the birds presented with the model eggs were divided into the categories listed (p. 306). Several birds showed more than one response and so the sum of the individual columns exceeds the total number of presentations in each case.

	Rolling[1]	Carrying	Lifting[2]	Billing	Nibbling	Poking	Ignoring	Total presentations
Empty egg	24	0	0	10	0	0	0	26
Filled shell	20	0	0	13	7	4[2]	3	27
Shell-with-lead	4	5	7	1	21	12	1	25
Real shell	5	13	2	4	16	3	3	27

[1] These categories include intention-rolling and intention-carrying respectively.
[2] Doubtful responses.

'Empty eggs' were always rolled in the same way as normal eggs. In Table 48 there is no significant difference in the treatment between 'Empty egg' and 'Filled shell', at least as far as the frequency of the responses is concerned. However, the latency of the response (i.e. the time between presentation and the start of the response) is significantly larger for 'Filled shell' than for 'Empty egg' ($p = 0.4\%$).[1]

For other conclusions we must bear in mind that some birds showed responses of both systems to one and the same model. As a basis of our statistical treatment, therefore, we took the numbers of birds which showed a certain response to some models but not to others, and compared these figures for each response with respect to different models.

The following conclusions can be drawn about rolling and intention-rolling. In all, 15 birds showed either rolling or intention-rolling

[1] All p-values in the remainder of this paper were computed according to the sign test.

to 'Empty egg' but not to the other models, not a single bird did the opposite, and 8 treated the models in roughly the same way. These figures are significant ($p < 0.1\%$).

Further, 13 birds showed rolling or intention-rolling to 'Filled shell' but to neither 'Shell-with-lead' nor 'Real shell', 1 did the opposite, and 12 responded in one of these two ways to all three models. These figures are also significant ($p < 0.1\%$).

As soon as a model had a thin edge or was hollow, therefore, it was less likely to be rolled in: this was independent of its weight. The interrupted outline did not affect the scores, though, as we have seen, the latency of the response is larger for 'Filled shell' than for 'Empty egg'.

Ten birds showed rolling or intention-rolling to both 'Empty egg' and 'Filled shell', but not to the two other models; no birds did the opposite; 7 did not treat these two categories differently, and 3 behaved similarly to part of the two categories. These differences, too, are significant ($p = 0.1\%$).

Billing gave very similar results. Eleven birds billed 'Empty egg' but not 'Shell-with-lead' nor 'Real shell', 4 did the opposite, and 10 treated the models similarly ($p = 5.9\%$). The number of birds which showed billing to 'Filled shell' but to neither 'Shell-with-lead' nor 'Real shell' was 13: 4 did the opposite, and 8 did not show this difference ($p = 2.5\%$). Finally, the number of birds which showed billing to either 'Filled shell' or 'Empty egg' but to neither 'Shell-with-lead' nor 'Real shell' was 21; 4 did the opposite, and 3 did not distinguish clearly either way ($p < 0.1\%$).

Table 49 summarises further information that can be extracted about the difference between 'Real shell' and 'Shell-with-lead'. The first line allows us to say that 'Real shell' elicited much more carrying, and the second shows that 'Shell-with-lead' was more often merely

Table 49. *Comparison of responses to 'real shell' and 'shell-with-lead'.*

This shows the numbers of birds which responded with each of the reactions listed in the first column to 'real shell' but not to 'shell-with-lead' (+), which did the opposite (−), or which did not differentiate (0) (p-values based on sign test).

	+	−	0	Significance
Carrying	7	0	5	$p = 0.8\%$
Lifting[1]	0	5	2	$p = 3.1\%$
Rolling	2	0	3	not significant
Nibbling	1	15	4	$p = <0.1\%$
Poking	0	10	2	$p = 0.1\%$

[1] This category includes intention-carrying.

310

lifted and then dropped. The fourth line shows that 'Shell-with-lead' very often gave rise to mere nibbling, while the fifth says the same about poking.

These data demonstrate that neither model elicited rolling, but that both elicited some or all of the removal sequence. This obviously consists of two parts, nibbling and carrying, and it is also obvious that the 'Shell-with-lead', while eliciting nibbling, stopped the real carrying which normally follows nibbling. In other words, the presence of the lead inhibited the action chain half-way. That this was due to the weight of the lead will become clear in the next experiment.

Experiment 3 This experiment was designed to test four hypotheses which we had developed as the results of experiment 2 took shape.

First, we had gained the impression that the 'Real shell' elicited removal mainly through having a thin edge, rather than through being hollow. We therefore offered an empty egg with a flange of shell glued on it at right angles to the surface. The flange measured 2 cm × 1·5 cm and could be considered flat. This model, which we called 'Flanged egg' (Fig. 106) was presented with the flange turned up and the longitudinal axis pointing radially. It thus offered a thin edge, but no hollow.

Second, to see whether the fact of a shell's being hollow had some effect, and also to have a control model in connection with 'Shell-with-lead', we filled a real shell with cottonwool so that it had very much the same appearance as the plaster-filled egg or 'Filled shell'. However, whereas the plaster neatly fitted the edge, the cottonwool (though concealing the hollowness) allowed the bird to see the thin edge as separate from it. This model we named 'Cottonwool shell' (Fig. 106).

Third, to see whether a serrated edge was more stimulating than a smooth one, we introduced two model eggs with similar-sized holes 40 mm from the blunt end. One had a smooth edge and the other was artificially serrated. These we called 'Smooth rim' and 'Notched rim' respectively.

Fourth, we thought that a real egg shell might owe part of its effectiveness to the amount of white it showed. A smooth-rim model (with a hole 40 mm from the blunt end) was therefore given a series of white triangles along the outside of the rim, so as to fake a serrated edge to the khaki exterior. This was known as 'Painted rim'.

With the 'Real shell' as a control, there were thus 6 models in this experiment. Each of these was presented 18, 19 or 20 times, the slight difference being due, as in the previous tests, to some of the series

Table 50. *Results of experiment 3*

This shows the numbers of birds which responded to the various model eggs with each of the reactions listed. The first two columns give the ultimate responses, irrespective of the incomplete movements which may have preceded them.

	Rolling[1]	Carrying[1]	Billing	Nibbling	Poking	Ignoring	Total presentations
Flanged egg	6	12	4	12	0	1	19
Cottonwool shell	6	13	5	13	0	1	20
Real shell	1	14	6	11	5	2	20
Smooth rim	15	2	5	5	2	1	18
Notched rim	9	9	7	6	1	0	19
Painted rim	12	8	12	4	5	0	21

[1] These categories include intention-rolling and intention-carrying respectively.

being prematurely ended by nest-relief. The results are given in Table 50.

Although the scores for rolling and intention-rolling in Table 50 are higher for 'Flanged egg' and 'Cottonwool shell' than for 'Real shell', the differences are not significant. 'Flanged egg' and 'Cottonwool shell' have not been compared in one series with 'Empty egg' and 'Filled shell' (Table 48), but neither of the latter models were carried even once; every single bird which responded to them did so by rolling. We must conclude, therefore, that 'Empty egg' and 'Filled shell' were treated like eggs (i.e. retrieved) and that 'Flanged egg' and 'Cottonwool shell' were treated like a real shell (i.e. removed), although we must leave open the question whether the birds distinguish to a slight extent between the different models within these groups. The main character to which the gulls respond by carrying must be one that is found in both 'Flanged egg' and 'Cottonwool shell' but absent from the other models. It follows that the thin rim is the important character and that neither hollowness nor broken outline, nor their combination (as found in 'Filled shell'), are sufficient to make the bird carry, although, as we have seen, 'Filled shell' elicited rolling with a greater delay than an egg does.

The high carrying score for 'Cottonwool shell' also shows that the low carrying response to 'Shell-with-lead' (Table 48) was due to its weight and not to the sight of something inside the shell.

Turning now to the characteristics of the shell's rim (the figures in the three lower rows of Table 50), we find that these scores are not significant either. However, because the differences between these models are relatively simple and straightforward, it is worth extracting fuller information from our protocols and counting all responses shown in all tests instead of merely listing (as in Table 50) whether or not a response occurred in any one test. Many birds responded more than once in a test; moreover, many birds showed in one and the same test elements of both rolling and carrying. By counting all these responses in every case, we can determine for each test an index which gives the ratio between the number of egg-rolling movements and the number of carrying responses—a 'retrieving-over-carrying index'.

By comparing these indices for the various models, we find that in 10 cases the index was higher for 'Smooth rim' than for 'Notched rim', in 6 the indices were equal for these two models, and 1 bird had a higher index for 'Notched rim' than for 'Smooth rim' (p = 0·6%). Thus 'Notched rim' elicited relatively more carrying than 'Smooth rim', which in turn elicited relatively more rolling.

This same index was higher for 'Smooth rim' than for 'Painted rim' in 8 cases, it was equal for both models in 8 more, and 1 bird showed the reversed response. These differences are also significant (p = 2%).

We can therefore conclude that both serrating the edge, as in 'Notched rim', and adding the white pattern of 'Painted shell' increased carrying and reduced retrieving.

Summary and Conclusions

Experiments were carried out at Ravenglass, Cumberland, in the summer of 1961, to determine how Black-headed Gulls (*Larus ridibundus*) distinguish between eggs and egg shells. These elicit entirely different sets of responses: eggs are rolled into the nest and brooded, while shells are picked up in the bill and removed. The following paragraphs summarise the conclusions reached.

The egg shell elicits removal because it differs from the intact egg in the following characteristics: it shows a thin edge; this edge is serrated; and it shows white. The 'decision' to remove it is already taken before the bird can have checked its weight. Neither the interruption of its outline nor its hollowness could be shown to contribute to removal, though both seem to exert an inhibiting influence on rolling.

In addition, it was shown that egg shell removal is a chain of acts: nibbling is elicited by visual stimuli; during nibbling the weight is checked and, if it does not grossly exceed that of a real shell, the object is carried; if it equals the weight of an egg or a chick, the chain is broken off. This prevents chicks from being carried away when they have hatched but not yet left the shell; whether other safeguards exist as well we cannot say.

Most of the population in which these responses were studied must have had previous breeding experience. Very little can be said, therefore, about the extent to which the responses are innate or could be the consequence of conditioning. Three one-year-old birds, which were tested with eggs and shells when they had not yet laid at all, rolled in an egg and removed a shell in a way indistinguishable from experienced birds (Tinbergen, Kruuk and Paillette, 3).

REFERENCES

1 SIEGEL, S. (1956). *Non-parametric Statistics for the Behavioural Sciences*, New York.
2 TINBERGEN, N., G. J. BROEKHUYSEN, F. FEEKES, J. C. W. HOUGHTON, H. KRUUK and E. SZULC (1962). 'Egg shell removal by the Black-headed Gull, *Larus ridibundus* L.: a behaviour component of camouflage', *Behaviour*, **19**, 74–117.
3 TINBERGEN, N., H. KRUUK and M. PAILLETTE (1962). 'Egg shell removal by the Black-headed Gull, *Laris ridibundus* L.: II. The effects of experience on the response to colour', *Bird Study*, **9**, 123–31.

(From the Department of Zoology, University of Oxford)

9
Food Hoarding by Foxes (*Vulpes vulpes*) L. (1935)[1]

Dedicated to Professor O. Koehler on the occasion of his 75th Birthday

During our studies of the behaviour of the black-headed gull (*Larus ridibundus* L.) in the sand dunes of the Ravenglass-Drigg Sanctuary, Cumberland, we gradually became aware of the importance of foxes as predators of these gulls in this particular area. Our observations began in 1959, when my friend Dr Ger Broekhuysen (Cape Town) and myself were struck by the number of newly fledged or near-fledged dead young we found in and around the gullery. They frequently showed signs of severe mauling, and fresh fox tracks could often be seen leading from one victim to the next. We soon realised that the Ravenglass dunes, with their extensive stretches of bare sand, presented exceptionally favourable opportunities for a study of the foxes' nocturnal activities through a systematic reading of their tracks. The fullest use was made of this opportunity in the years 1961, 1962 and 1963 (Kruuk, 1). My own observations, though somewhat intermittent, cover an even longer period, and some of my findings, especially those of the years 1963 and 1964, are new and supplement Kruuk's observations.

In the beginning, we were mainly interested in the gulls and their modes of defence against foxes and other enemies. But soon we became fascinated by the behaviour of the foxes themselves.

Occasionally I. J. Patterson, myself, and Kruuk in particular, observed the foxes directly but most of the findings reported here are based on an interpretation of tracks (Fig. 107). The illustrations supporting this paper[2] will serve to give some idea of the reliability of this

---

[1] Part of these observations were done together with Dr Monika Impekoven (Basle) and Dr Dierk Franck (Hamburg); I also owe Michael Norton-Griffiths some valuable data concerning our 'egg line experiment'.

[2] Of the 41 photographs illustrating the original paper, only a few can be reproduced here.

Fig. 107. Track of fox on his way to gulleries.

ancient method. In learning to read the tracks we were greatly helped by the papers by Tembrock (3) and, in particular, Murie (2).

Early mornings, especially mornings following dry, quiet nights, were best for observation and documentation. Dew, or light white frost over dry sand produced the best conditions. It is amazing how much one can detect, when one looks (or takes photographs) against the light while the sun is still low.

The foxes gradually increased in numbers from 1960 to 1964, when, in February, 6 of the 7 foxes then inhabiting our peninsula (which was *c*. 1 × 3 km in size) were killed. In 1962 the gulleries were regularly visited by 4 adult foxes, and occasionally by a fifth; in addition we counted 12 cubs (Kruuk, 1, p. 37).

Analysis of faeces showed that in winter the foxes subsisted mainly on rabbits but also caught small rodents and birds. When the gulls arrived at their breeding grounds some of the foxes began to concentrate increasingly on gulls and their broods. From the middle of March onwards, the gulls spent part of each morning on the dunes. At first they visited them only in the very early mornings and were very wary but their visits gradually lasted longer, and their behaviour became more confident. Until the gulls started to lay eggs, the foxes spent the rest of the days foraging on ploughed fields and on the banks of the river. At night-fall the gulls retired to the wide open spaces of the beach to roost, and there the first victims of the season were invariably found. However, the foxes needed exceptionally dark nights (new moon and solid cloud cover) for success on these roosting places. No doubt a few of the gulls that were killed in these dark nights were eaten on the spot or carried to the foxes' earths, but the majority were just left lying where they had been killed. As soon as the gulls started laying and nests became occupied at night, more and more attacks were made on the actual breeding sites, and during this period exceptional darkness was not necessary. The foxes killed a great many birds which were sitting on their nests and clutches and ate great quantities of eggs, but even then they still caught many rabbits as well. Later in the season massive numbers of young birds fell victim to the foxes. During this period, which lasted until the gulls left in July, the foxes' behaviour was the same as described before: the majority of the birds that were killed, and of the eggs that were taken, were not eaten. Over a period of three years, Kruuk and his collaborators found no fewer than 1,449 adult black-headed gulls that had been killed and abandoned. The number of chicks and eggs to which this had happened could not be determined. It must have run into many thousands.

Kruuk found that the black-headed gulls possessed very effective means of defence against most of their enemies but were almost completely defenceless against foxes. While I believe (without as yet being able to prove this) that their habit of scattered nesting, resulting from the territorial system, provide a degree of protection against predators, it seems clear that the dune habitat is far from optimal for black-headed gulls. There is considerable evidence that black-headed gulls, in common with other 'masked' gulls, are not really coast dwellers, but are most at home on freshwater lakes inland. But draining of inland marshes has increasingly forced them to settle in coastal dunes. Though foxes are good swimmers, they apparently do not like entering the water and it seems probable that in the typical freshwater habitat, the danger from foxes is a good deal less. It is, moreover, likely that man, through his agricultural activities, especially ploughing, has contributed to the build-up of gull populations in certain areas and so has led to the large breeding colonies we observe today. And finally, there is no doubt that a density of foxes such as we have had at Ravenglass in recent years, is abnormal—attributable, presumably, to the fact that they were not controlled. It has to be realised that man centuries ago took over the role of predator from wolves and large birds of prey in virtually the whole of Europe, but has withdrawn from it again recently in isolated areas such as modern sanctuaries. Thus, ironically, we have now to acknowledge that our aim of studying black-headed gulls in their natural habitat could not be fully reached in Ravenglass, which must be considered slightly 'unnatural'.

Such mass killings as we had occasion to observe (in one exceptionally favourable night in March 1962 we found 230 freshly killed adults in the gullery) might appear senseless. But I rather believe that they are induced by the abnormal habitat. For one thing, they occur only under certain well-defined, infrequent conditions, e.g. in chicken houses, or in large and dense dune colonies like ours in Ravenglass. It is furthermore quite possible that a hunting animal such as the fox can hope to be successful (even on favourable nights) only if it possesses an unquenchable urge to kill on sight, to kill immediately whenever opportunity offers. It is quite possible that it cannot afford to curb its instant reaction towards 'moving-prey-within-reach' under any circumstances whatsoever. And finally, foxes are in the habit of burying part of their prey which might well be a method of preserving food for lean times. Of course, before we can ascribe this function to the hoarding habit we must know whether foxes actually use the prey they have hidden. This problem—the food catching habit of foxes and its survival value—is the subject of the present paper.

Foxes cache mostly eggs, but occasionally also other prey. A fox wanting to hide its prey digs a small scrape with the front paws, places the egg or dead animal into it, and covers it quickly (as dogs do) by shovelling sand or plant debris over it with its snout (Fig. 108, p. 320). The whole operation takes only a few seconds. Alternatively it pushes its prey (especially rabbits; never eggs) deep into some patch of dense vegetation (marram grass, or even nettles). This caching habit might well be adaptive in two respects—provided of course that the hoards are really used at a later date. For one thing, caching prevents the finding of the prey by the ubiquitous carrion crows which occasionally follow the foxes on their predatory expeditions; prey that is not carefully hidden will more often than not have been eaten by crows early the following morning. Secondly, their food hoarding behaviour shows a peculiarity, the significance of which I shall discuss later in this paper: foxes are 'scatter hoarders'; they never hide more than one prey in any one place. This is particularly noticeable with clutches of eggs. A fox picks up one egg at a time, carries it in its mouth for a distance of 5–10 m, buries it in the manner described and returns to the nest. It will then pick up another egg and bury it at a similar distance, *but in another direction*, and return once more to repeat the process with a third egg, and so on. The eggs from one gull nest (usually three) are therefore always hidden quite far apart.

During the early years of our investigation we never succeeded in finding evidence that the foxes had actually returned to dig up their hoarded prey although we checked known caches frequently and with great care. Our failure could of course have been due to an unavoidable methodological difficulty; obviously, the only caches we could check were those which we had previously found, and these would always be contaminated by our scent, even though we could often locate them from some 10 m, and so avoid coming nearer. However, though the possibility that the foxes avoided caches which had been visited by us could not be excluded, this was not very probable, for the Ravenglass foxes were remarkably indifferent to us. For instance, they were not even temporarily driven away by our occasional patrolling of the gullery by night, and they sometimes carried away dead gulls that we had handled. Moreover they frequently approached to within 50 m of our camp, and sometimes even dug up our rubbish dump.

Not until 1963 did I realise that we should have to continue our observations much longer than we had done hitherto. Up to that year we had ceased our work and so our regular checks as soon as the gulls and their young had left the locality, i.e. about the middle of

a

b

Fig. 108. (a) a freshly made egg-'cache'; marks of feet, snout and whiskers of fox clearly visible. (b) a fox 'cache' dug up in summer; arrow indicates the broken shell of an egg of a Black-headed Gull.

July. I therefore decided to return to Ravenglass towards the end of July. Daily checks of those parts of the colony site where many cached eggs had been found earlier in the year now showed clearly that the foxes continued to visit these localities regularly, and that they did actually dig up the eggs that had been lying buried for appr. two months. I found a total of 30 eggs that had been dug up and eaten on the spot: of these, 27 were eggs of black-headed gulls, and 3 were of the merganser (*Mergus serrator* L.), another dune-breeding species which suffered severely from fox predation. But freshly dug up eggs were only found in the last week of July. Later in the season the foxes did not visit these parts of the colony particularly often, but wandered all over the peninsula, concentrating once more on rabbits and other prey. It is probable that by then they had used up all the eggs they had cached. After all, nearly a fortnight had passed between the departure of the gulls, around July 10th, and my return to the site, and it seemed natural to expect that the foxes would start visiting their hoards as soon as food became scarce through the sudden disappearance of the gulls. Tracks that were still visible but partly obliterated also pointed to my having arrived towards the end of that particular period.

These observations demonstrated that the food caching habit of the foxes does have survival value, and also that it presents more adapted features than I had originally assumed. For not only is the act of caching effective: the timing of retrieval is too; the foxes would not start digging up their hidden prey until the need arose. And yet we might easily have concluded from our original observations in May and June—which had been done with considerable care—that the caching of food was a senseless habit. It is by no means superfluous to stress this, for such negative assertions, which sound wellfounded but in reality are uncritical, are only too often put forward in discussions of survival value.

Careful scrutiny of the foxes' tracks showed some further unexpected features. We had realised for quite a time that the foxes knew their hunting grounds in great detail, and that, for instance, they would always follow the same route between their earths and their hunting areas. Sometimes we found what almost amounted to a highway along the seashore (cf. Kruuk, 1); they also made use of very definite 'passes' to get into the dunes, and their tracks quite often stretched in very nearly straight lines across wide areas of sand in the direction of the gulleries. If, as sometimes happened, they were attracted by some smell, they might trot, say, 10 m upwind at right angles to their main direction in order to dig up a half-buried carcass or the like, but very soon they would resume their former course.

Individual foxes (we could sometimes recognise individuals by their tracks) had their own preferred hunting grounds and fought other foxes if they encountered them there. A fox entering a strange hunting ground changed its behaviour: instead of running in a straight line across the open sands it would try to keep under cover and would move hesitatingly and follow an irregular course.

There was therefore nothing surprising in the fact that the tracks I found in late July 1963 led in a straight line to those parts of the gullery where eggs were buried. But once there the behaviour of the foxes had changed very strikingly. The tracks criss-crossed the sand, and the individual footprints were very close together, a sure sign that the fox in question must have wandered about slowly. The tracks, moreover, ran from one landmark to the next (bushes of marram grass, hillocks, large pieces of driftwood, etc.) and showed that the fox had stopped at each of them. I dug up the ground at more than 50 places where the fox had stopped (and presumably sniffed) but had not dug. On none of these sites could an egg be found. However, in all 30 places where the animal had dug in the sand, I found broken egg shells (Fig. 108, p. 321). It was clear how the foxes go about retrieving their hoards: they do *not* remember individual hiding places; they remember only in what part of the gullery, roughly, they have cached their prey. During their 'retrieval-foraging' they are at first guided by visual cues, just as they had been during the caching itself. They walk up to certain landmarks and, once there, they presumably smell whether or not an egg is buried in the vicinity. In this connection it is perhaps significant that none of the eggs dug up had been covered with more than 5 cm of sand. We must assume that those eggs covered by deeper drifts are lost. Some tracks enabled me to infer the distance from which the fox had been able to smell a buried egg. For occasionally I came to an area where the fox had not (to my knowledge) buried any eggs—in the centre of a bare patch where there were no landmarks (nothing, in fact, but ever-changing ripples in the sand). Here tracks showed that a fox, running in a straight line, had suddenly stopped in its tracks, made a sharp turn upwind, and dug up an egg. As a rule, these sideways moves covered no more than 50 cm, but in one instance the distance was 3 m (admittedly the egg concerned turned out to be a severely damaged (and pretty 'high') merganser egg).

However convincing this observation appeared to be, an experiment was required. Consequently we now decided to test whether foxes can find eggs that they have not themselves buried, and of which they could not possibly remember even the general location. In May 1964 therefore, our guest workers, Dr Monica Impekoven from Basle, Dr Dierck Franck from Hamburg, and I myself, buried a number of eggs, imitating the foxes' manner as best we could, in areas habitually *traversed* by foxes on their way to and from their hunting grounds, but away from places they themselves used to bury eggs. We 'cached' 100 fresh hens' eggs in a bare strip of sand across the foxes' habitual approaches to the general area of the gulleries. The eggs were spaced 10 m apart, and alternating with them we made in all 100 similar scrapes which we covered up without putting an egg inside. Each of these 200 spots was marked with a numbered wooden peg inserted appr. 2 m away; the entire 'egg line' therefore was 1 km long.

We now checked this line daily from May 13th to August 28th (when the last of us had to leave Ravenglass). The checks were made early each morning, except on a few very wet and windy days (when we should not have found any readable tracks anyway), and we took care to follow the line of markers, avoiding our 'caches'.

Unfortunately this experiment was carried out under extremely unfavourable circumstances. The reasons for this were twofold. In the first place the gamekeeper, as already mentioned, had made strenuous efforts in February 1964 to reduce the fox population to a number tolerable from the point of view of the gullery. In this he had succeeded only too well, killing all foxes but one. This seventh fox, a large, strong dog fox, remained in the area but visited our gullery only about once a week, was moreover not a very great gull killer; he was a keen hunter of rabbits which, though they had been scarce in March, had multiplied rapidly that year. Altogether we could only find some 25 eggs he had buried himself. The second difficulty was the weather. In 1964 heavy spring gales were exceptionally frequent and shifted enormous quantities of sand so that a large proportion of our eggs came to be buried far too deeply.

Nevertheless, this particular fox, and another one that turned up sporadically, dug up a total of 8 eggs in the course of the summer (Fig. 109a, p.326): 1 on June 15th, 2 on June 30th, 1 on July 15th, 2 on July 16th, and 2 on August 21st.[1] Neither fox took any interest in our fake caches.

Of course the number is small. But it should be remembered that we deliberately did not bury our eggs in the hunting ground itself. And in addition we are in a position to quote another dozen or so cases in which a fox dug up eggs presumably cached by another fox, or rabbits and other carcasses also not buried by itself. Once we

[1] Three more eggs were found dug up and eaten on a visit in December. Next spring all but 5 had disappeared—all of these were lying under at least 10 cm of sand. Already by December they had a curious solid consistency and to us gave off no scent at all.

found the tracks of both foxes involved. We could see how the 'rightful owner' had returned to its empty hiding place the day after a robber had unearthed a rabbit hidden there and had reburied it 30 m away. Another time, towards the end of July, I found a place where a fox had unearthed an old and very smelly merganser egg which had recently been buried. Realising that this egg must have been stolen from another cache, I retraced the fox tracks, and arrived at a place 600 m away where the fox had made the typical turn into the wind already described, and had gone on until it had found the egg. Later we confirmed that this digging up and reburying of eggs was especially characteristic of foxes that had entered the hunting ground of another fox. The 'outside' fox that had visited our egg line only once, on July 16th, had behaved in this fashion; unlike the resident fox, it had not eaten the two eggs on the spot, but had carried them away. One we discovered after a laborious search, the other was carried so far into a grass-covered area that we lost the fox's track. Cases where foxes had found carcasses completely buried by drifting sand are additional evidence for our assumption that foxes find hidden hoards by smell.

Taking all this evidence together there seems therefore no doubt that the 'caching' does preserve at least part of the surplus food obtained in times of superabundance; that it is done in specific areas; that these areas are remembered; that within them the fox finds individual caches by scent; and that this stimulus is powerful enough to make a fox dig up even eggs buried by others.

In the course of our daily early-morning checks we incidentally made another highly interesting observation: we found that hedgehogs had dug up no fewer than 12 of our eggs (Fig. 109b, p. 327). Various indications showed that hedgehogs, like foxes, can smell eggs, but that they can find them only when they stumble upon them accidentally. The furthest distance from which they can discover an egg covered by 3 cm of sand is approximately 50 cm. We knew of course that hedgehogs were confirmed robbers of gulls' eggs (Kruuk, 1); but it was news to us that they were in the habit of plundering the hoards of foxes. We have never found evidence to suggest that hedgehogs bury eggs themselves.

These 'thefts' by hedgehogs, and also the thefts of foxes from each other, made me consider once more the survival value of scatter hoarding as distinct from 'clump hoarding'. The problem is not simply one of overall risk of loss of food. We know on the one hand that a hedgehog that finds 3 eggs in one place, say in a gull's nest, will frequently eat the lot. Kruuk found that hedgehogs offered no other food ate on an average 5·3 eggs a night and destroyed 1 more.

a

If a fox were to put all eggs from one gull's nest into one cache, or even were to bury them close together, they would all be lost to him should a hedgehog find the hoard. On the other hand, the number of hoards would obviously only be a third of the number of single-egg hoards, and, seeing that the discovery range of hedgehogs is so small, the risk of discovery would be correspondingly reduced to approximately a third. (We have in fact never had a case where a hedgehog had found two neighbouring eggs in our egg line in one night, though on two occasions two eggs, buried several hundred metres apart, were eaten by hedgehogs in one night.)

From this point of view the advantages and disadvantages of scatter hoarding might seem to cancel each other out. But we cannot say this with certainty until we have more information on the distance from which a hedgehog can detect a group of 3 eggs, how thoroughly and how specifically it continues to search for eggs after the first find, the time of the year when hedgehogs and foxes begin to dig up egg caches, the rate of searching of both species, and so on.

However I believe we *can* say this: even assuming that in the long run egg losses would not differ between the two strategies (i.e. 'clumping' and 'scattering') or assuming that they might even be

b

Fig. 109. (*a*): where a fox has dug up an egg of our 'egg line'; arrow
1 indicates a footprint, arrow 2 part of the egg shell. (*b*): tracks of
a hedgehog circling round an egg 'cache' before robbing it; arrow
indicates egg shell.

slightly greater where the hoards were scattered, the danger of
occasional *very high* losses, which could really be critical for the fox,
would be greater if he did not, as described, spread out his finds so
systematically.

We may be justified in extrapolating to say that in general, the
smaller the number of caches for a given number of eggs, the greater
the chance that occasionally *all* eggs will be found. That this could, in
lean periods, put the fox in a critical situation, is clear when one
remembers that a cache, once opened, is usually lost, if only because
the far-roaming and extremely keen-eyed carrion crows would finish
it. With scatter hoarding, the risk of a really serious loss is smaller
than with larder hoarding—the fox confirms the validity of the old
saying that one should not put all one's eggs into one basket.

327

REFERENCES

1 KRUUK, H. (1964). 'Predators and anti-predator behaviour of the Blackheaded Gull (*Larus ridibundus* L.)', *Behaviour Suppl.*, **11**, 1–130.
2 MURIE, A. (1936). 'Following fox trails', *Misc. Publ. Un. Mich. Mus. Zool.*, **32**.
3 TEMBROCK, G. (1957). 'Zur Ethologie des Rotfuchses (*V. vulpes* L.) unter besonderer Berücksichtigung der Fortpflanzung', *Der Zool. Garten* (*N.F.*), **23**, 289–532.

(From the Department of Zoology, Oxford University)

10
An Experiment on Spacing-out as a Defence Against Predation (1967)

This study is based on the hypothesis that certain predators exert a pressure on individuals even of well-camouflaged prey species to live well spaced-out; more precisely, to live at interindividual distances which greatly exceed the distance from which predators usually detect them directly. This hypothesis occurred to the senior author in the course of studies, partly by himself, partly by his co-workers (De Ruiter, **9, 10**; Tinbergen, **12, 13, 14, 15**; Tinbergen *et al.*, **16**; Patterson, **8**), and in connection with certain data in the literature. The evidence leading to the hypothesis is of various types; each kind of evidence is in itself admittedly incomplete or even tentative, but in their totality the data were considered suggestive enough to warrant the present study.

It is of course well realised that camouflage, in the sense of colour patterns hampering the detection of a prey by predators that hunt by sight, is fully effective only if accompanied by certain specific types of behaviour. Those generally recognised, and experimentally checked at least in some cases, are: (1) immobility at least by day (e.g. De Ruiter, **9**); (2) living on the background which matches the animal's colouration (e.g. Kettlewell, **4, 5**); and (3) adopting a position which provides maximum concealing effect (e.g. Cott, **1**; De Ruiter, **10**).

We suggest that living well spaced-out (in the sense defined above) is another behavioural correlate of camouflage.

The evidence which gave rise to the hypothesis can be summarised as follows:

1. While no camouflaged animal is safe from detection, the distance at which even keen-eyed predators detect them is usually remarkably short. No systematic measurements are available, but the statement describes the general experience of those who have observed predation of camouflaged animals either in the field or in laboratory tests.

2. Although, again, exact measurements are rare (and would in many cases be very difficult to obtain), it is the general experience of naturalists that the majority of camouflaged animals live well spaced-out; to be more exact, it is the discrepancy between the 'Direct Detection Distance' and the actually observed interindividual distances in populations of many camouflaged animals which strikes the observer who has some experience of both, and which makes him suspect the existence of a hitherto undetected pressure favouring this degree of spacing-out.

3. It is of course true that many animals have mechanisms which ensure spacing-out, and that the survival value of these mechanisms often has nothing to do with predation. The most common advantage is perhaps the reservation of an adequate food supply (see e.g. Lack, 7); another is the reservation of a suitable breeding site. With many camouflaged animals it looks as if none of the known or suspected advantages apply or are critical, and thus, by elimination, one begins to suspect that they must be subject to another type of pressure.

4. This suspicion is strengthened when one considers the ways in which different species achieve spacing-out. The mechanisms used vary greatly; for instance, egg-laying females of some Sphyngid moths whose larvae are camouflaged, fly for a considerable distance between the laying of two successive eggs (Tinbergen, 13); and although the eggs of the Peppered Moth are laid in a clump, the young larvae scatter by means of wind dispersal before they develop their well-known twig mimicry (Kettlewell, pers. comm.). This diversity of means achieving the same end suggests adaptive convergence.

5. The hunting behaviour of some predators, particularly birds, seems to provide the clue. There are indications that at least many birds react to the discovery of a prey item by an intensified effort in the area round such a first find—they search round the place where the original prey was found. It is probable that such 'area-restricted searching' is, in addition, restricted with respect to the type of food—that the predator searches specifically for a type of food similar to that of the original prey. This 'hunting by searching image' was first suggested by L. Tinbergen to account for the fact that the toll levied by Tits of a gradually increasing population of certain insects was at low densities below the percentage to be expected on the basis of random encounters, but rose above this expected level when the prey species became more numerous. While we believe that restriction of the search to a certain area in such cases occurs together with a restriction of the type of prey sought for, we should like to empha-

sise that 'area-restricted searching' alone would already lead to an 'Effective Detection Distance' which would be larger than the'Direct Detection Distance', and so would put an extra premium on spacing-out by the prey species. If the formation of a searching image were to come into play as well, then the pressure for spacing-out would be stronger still.

Our study, based upon these considerations, aimed at testing whether the risk run by camouflaged prey populations of different densities, exposed to predators who had been alerted by one prey sample, would vary with the density according to the hypothesis. The work is the first step in the study of this problem; further work by H. J. Croze (on which he reported briefly at the International Ethological Conference in Zürich, 1965) is in progress (2).

Method

Our experiment was designed to test whether, once a predator had found one camouflaged prey, the mortality among a population of other, similar prey items would be greater at higher than at lower densities: in other words, whether predators can under certain circumstances penalise crowding. We also hoped to test, by direct observations of the predators at work, whether they went in a straight course from one prey item to the next, or whether, when faced with the more widely scattered populations, a non-oriented, searching phase would precede the straight approach to each prey item. This, if true, would show that the Effective Detection Distance was indeed larger than the Direct Detection Distance.

Wild Carrion Crows (*Corvus corone* L.), who had proved to be extremely keen-eyed predators in previous experiments demonstrating the effectiveness of camouflage of eggs of Black-headed Gulls (Tinbergen *et al.*, 16) were used as predators, and well-camouflaged eggs were presented in different densities as prey. The experiments were done in 1964 in the same area as those done on camouflage, viz. in valleys in the sand dune peninsula near Ravenglass, Cumberland.

We used Hens' eggs, painted in a standard pattern resembling that of the eggs of Black-headed Gulls and of many other ground-breeding Laro-Limicolae. We have indications (although no strict experimental proof as yet) that the irregularity of the blotching pattern (in size, shape, placing and hue of individual dots) is of great importance to the effectiveness of the camouflage, and also that one or two straws loosely laid over an egg makes it even less conspicuous. For these reasons we took great care to paint our eggs in a natural pattern which at the same time we standardised as much as possible. To camouflage them even better, we dug them in until approximately half their side showed and covered each egg with a few straws. Such eggs were extremely difficult to see and, as our data will show, were often overlooked by the Crows.

The eggs were laid out (in the absence of the Crows) in plots of

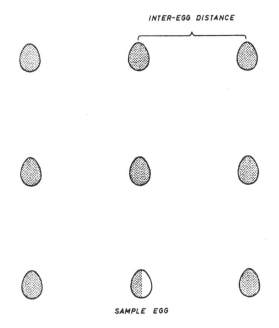

Fig. 110. Arrangement of a plot of one sample egg (bottom row, centre) and eight concealed eggs.

different density in areas covered densely with moss, lichens and other low vegetation, with occasional tufts of Marram Grass; we then retreated to an observation hide at approximately 150 m from the plots, waiting for the Crows to arrive. For each test 2 groups of 9 eggs each were laid out in squares of 3 × 3 eggs, with a fixed distance between adjacent eggs of the same field (Fig. 110). The 'inter-egg distance' differed for the two squares offered in each test: those first used were 50 and 800 cm respectively; in other tests the difference in density was less. The densest egg field covered a surface area of 1 sq. m; the least dense field one of 256 sq. m. One egg in each square was not dug in and was unpainted on one side so as to be more conspicuous than the others (Fig. 110). This 'sample egg' served to attract the predators to each field. In our predation tables these sample eggs are not included; they refer to the 8 camouflaged eggs that were available in each plot. There was always a distance of at least 30 m between the two plots offered in each test; this proved sufficient to prevent the predators from 'overflowing' from one of the two plots into the other; while they usually arrived flying, they walked once they had alighted in a plot, and they had to fly and search anew to discover the sample egg of the other plot. We have notes on walking and flying of 24 tests in which the birds were undisturbed during the entire test; in 20 cases the Crows flew up when leaving one plot; in the other 4 tests one Crow walked from one to the other. In all cases it was clear from the behaviour of the Crows, which was initially not oriented to the second sample egg, that they did not discover it until after they had left the first plot.

In a further 5 out of 8 tests, in which the birds were disturbed, they left without taking the sample egg, or indeed any egg of the second plot. Almost all tests were paired to the extent that the same two densities were offered twice in succession with the position of the scattered (S) and crowded (C) eggs interchanged. Between pairs of tests the location of the two plots within the total area of approximately 5,000 sq. m were changed.

Once the Crows were used to finding food somewhere in our 5,000 sq. m area, they visited this several times a day. Consequently we rarely had to wait longer than two hours before they arrived. They used to approach flying some 5–10 m above the ground, looking down, and would suddenly stall and alight near an egg, which was then approached walking. We have reliable notes about the eggs discovered first in this way in 66 plots. In all but 6 of these it was the sample egg which was found first, and we must therefore conclude that it was usually the sample egg which started off the search. S- and C-sample eggs had equal chances of being discovered: in 25 undisturbed tests the S-eggs were found first on 12 occasions, the C-eggs on 7 occasions, while C and S eggs were found simultaneously (2 Crows being involved) on 6 occasions. The first egg was usually eaten on the spot and the subsequent search was conducted on foot. It was a fortunate circumstance for the purpose of our tests that the Crows did not give up the search when they had eaten their fill. While they consumed no more than 1 or 2 eggs per test, the Crows went on taking eggs, which they carried off in flight and buried at distances varying from 30 to 300 m. After having quickly buried an egg[1] they would return to the same plot and continue their search.

[1] They retrieved many of these buried eggs days or even weeks later. We once saw a Crow fly directly to such a 'cache' 4 days after it had buried the egg (which we had long since removed).

Table 51

Test series	Date	Place	Birds	Inter-egg distance in metres		Number taken		Number of tests	Sign. level for diff.
1	9–11.3	Big Valley	pair A	0·5	8	24	5	3	$p < 0.001$
2	12–16.3	Big Valley	pair A	1	4	48	44	6	$p > 0.1$
3	17–18.3	Big Valley	pair A	4	8	16	16	2	$p < 0.001$
4	21–29.3	Lichen Valley	pair B	0·5	8	23	9	3	$p < 0.001$
5	21.3–2.4	Lichen Valley	several pairs	0·5	8	27	14	4+4[1]	$0.1 < p < 0.2$
6	4.4–10.4	Big Valley	pair A	2	8	52	25	7	$p < 0.001$
7	4.4–12.4	Lichen Valley	pair B	2	8	30	6	6	$p < 0.001$

(χ^2 tests, 1-tailed)

[1] One plot visited in each test.

The Crows usually came in pairs. The male sometimes hunted alone with the female following him, often the mates hunted independently, but one bird could, and we were certain did, always notice the discovery of a sample egg by the other. From peculiarities in their behaviour and appearance we are convinced that all experiments in 'Big Valley' (see Table 51) involved the same pair; in 'Lichen Valley' another pair was operating, which was sometimes joined (and interfered with) by other pairs. During their search the Crows would visit the two plots in succession; sometimes one Crow would feed in one plot, its partner in the other; on still other occasions (usually when they were disturbed) they would leave before having visited the second plot.

We concluded each test when 10 minutes had elapsed after the Crows had given up the search, i.e. either flown away out of sight, or flown or walked to areas well outside the 5,000 sq. m test area. Almost all undisturbed tests ended with a period of unrewarded searching.

After a test we removed all eggs, other than those buried by the Crows themselves (which, since we concentrated on their behaviour in the egg plots, we often failed to locate). No more than 2 tests were done on any one day, and these were always separated by an interval of at least 2 hours.

The data recorded were: (a) the number of eggs taken in any plot visited; (b) the time interval between abandoning one egg and finding the next; (c) how long, after having dealt with the last egg, the Crows searched until they left the plot; and (d) the total time spent in each plot. We also tried to get an idea of the Direct Detection Distance by observing at what distance a Crow's random walking changed into a straight approach, but under the conditions of the test this was very difficult to do.

Results

(a) *Numbers of eggs taken*   Table 51 summarises the eggs taken in 7 test series. Only in series 5 was more than one pair involved; this series includes the data for 8 tests in which one plot had been visited and abandoned before the arrival of a second pair disturbed the test; it so happened that 4 of these were S-plots and the other 4 C-plots. In series 1 and 4 (50 against 800 cm) significantly more eggs were found in the C-plot than in the S-plot. The difference is indeed striking; of 48 crowded eggs presented in series 1 and 4, all but one were taken, whereas of the 48 scattered eggs presented in these same series, 34 were not found. In series 5 (also 50 against 800 cm) the differences, while still indicated, are not significant. In test series 2 the inter-egg distances were moved closer together: 100 and 400 cm respectively. The predation was not significantly different for C and S.

Series 6 and 7 show that when the C-eggs were 200 cm apart and the S-eggs 800 cm, the C-eggs were significantly more vulnerable. In series 3 (400 against 800 cm) the Crows took all S-eggs as well as all C-eggs. This was the last series in a continuous run of 10 days' testing with pair A, whose efficiency seemed to increase with experience.

Thus, under these conditions, crowding was usually penalised, sometimes severely.

(*b*) *Searching time per egg* Table 52 gives the mean time spent searching after dealing with one egg and finding the next. We do not have good records for series 1, but those of all other series show that

Table 52

Test series	Inter-egg distance in m		Number of tests	Mean time to find one egg		Sign. level for diff.
	C	S		C	S	
1	0·5	8	3	not measured		
2	1	4	6	44″	1′30″	p < 0·01
3	4	8	2	1′06″	1′51″	
4 + 5	0·5	8	3 + 4	1′08″	2′15″	p < 0·032
6	2	8	7	56″	2′26″	p < 0·001
7	2	8	6	3′10″	3′33″	p < 0·1
						(Mann-Whitney test, 1-tailed)

it took more time to find an S-egg than to find a C-egg; in series 2, 4 + 5, and 6, significance is at, below, or not far from the 1% level. This means that the reward rate per time unit was indeed higher for the crowded eggs than for the scattered eggs.

(*c*) *Searching after last egg* The relevant data are given in Table 53. In all but one series for which we collected data the Crows, before giving up, searched longer in the S-plot than in the C-plot; in series 4 + 5 this difference is significant at about the 4% level. In series 7 the difference is reversed but not significantly so. These data show that the lower score for the S-eggs was not due to the birds giving up more readily in the S-areas.

(*d*) *Total time spent in each plot* Table 54 finally summarises what we could call the total effort. While the total time of exposure was

Table 53

Test series	Inter-egg distance in m		Number of tests	Searching time after last egg		Sign. level for diff.
	C	S		C	S	
1	0·5	8	3	not measured		
2	1	4	6	3′10″	3′20″	p > 0·1
3	4	8	2	2′15″	4′15″	
4 + 5	0·5	8	3 + 4	1′47″	4′17″	p < 0·038
6	2	8	7	3′26″	4′	p > 0·1
7	2	8	6	2′35″	1′20″	p > 0·1 (Wilcoxon test 2-tailed)

Table 54

Test series	Inter-egg distance in m		Number of tests	Total time spent in area		Sign. level for diff.
	C	S		C	S	
1	0·5	8	3	16′	11′	p > 0·1
2	1	4	6	9′	14′10″	p < 0·05
3	4	8	2	12′	20′	
4 + 5	0·5	8	3 + 4	8′38″	13′56″	p > 0·1
6	2	8	7	10′34″	13′30″	p > 0·1
7	2	8	6	14′15″	10′55″	p > 0·1 (Wilcoxon test, 2-tailed Mann-Whitney test, 2-tailed)

always equal for the two plots of any one test, the time actually spent by the Crows in each plot (the 'effective exposure') differed. In all but two series the total effort was larger for the S-plots than for the C-plots; in series 2 this difference is significant at the 5% level, in 4 + 5 it approaches significance. In series 1 and 7 more time was spent in the C-plot.

All data in Tables 52, 53 and 54 therefore show that the lower mortality of the S-eggs shown in Table 51 was not due to a lower effort on the part of the predators; rather the contrary was the case.

There are some indications that our hypothesis, although confirmed to the extent that the mortality among crowded eggs was higher than among scattered eggs, was too simple. The fact that the Crows searched longer in the S-areas before giving up suggests that they had registered the fact that a longer search could pay in these plots; it might be significant that in the first tests (series 1) the Crows actually spent longer in the C-plot than in the S-plot. It looks as if, when feeding on one prey for a long time, the Crows learned to assess their usual density. This of course would be a consequence of our method, and might not happen in nature.

Discussion

Some caution is required in formulating exactly what the above experiments do and what they do not demonstrate. While they prove the existence of a potential selection pressure of the kind required by our hypothesis, they do not give information about the effectiveness of this pressure under natural conditions.

1. There is of course no doubt that under the conditions of the tests the relatively more crowded prey was penalised more heavily than the relatively less crowded prey. We conclude that this was due to the Crows actually failing to find the eggs which they did not take, in spite of attempting to find food. This conclusion is based on our interpretation of the Crows' behaviour as searching, and this rests on the following criteria: the course they followed while walking over and near the plots was usually irregular, even in parts of the tests with 50 cm inter-egg distance, and this behaviour usually ended abruptly either in a short straight walk which resulted in the taking of an egg, or, at the end of each test, in departure. It was impossible to collect accurate data on the lengths of those straight approaches, but it was obvious that the Direct Detection Distances, while naturally varying, were of the order of 50 cm and not infrequently even shorter. The fact that the Crows usually searched before they found an egg, and that by doing so they managed to find eggs 400 and even 800 cm away from the egg found previously shows that the range over which they constituted a risk to the eggs (which we could call the 'Effective Detection Distance') was considerably larger than the Direct Detection Distance. The search was indeed area-restricted —when the Crows gave up the search they almost invariably flew away to other feeding grounds at least 500 m away.

We should like to emphasise that the *fact* that the Crows penalised the relatively more crowded eggs more heavily than the relatively less crowded eggs must be considered separately from our *interpretation* of the way in which the Crows' behaviour resulted in this effect.

We conclude that the area-restricted searching of a predator can exert a pressure on a prey species which tends to favour larger inter-individual distances than the Direct Detection Distance of the predator would require. Although we have indications that the Crows, while on our plots, ignored food such as rabbit carcasses which they would take on other occasions, we have to leave open the question

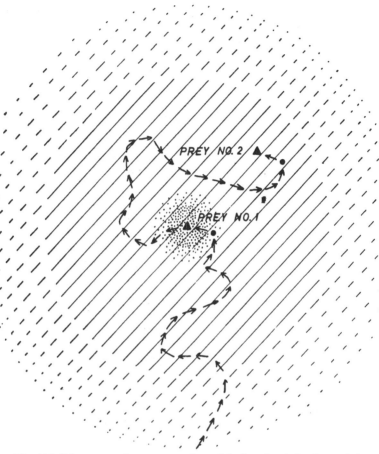

Fig. 111. Diagrammatic representation of the hunting behaviour of the Carrion Crows as observed, and the consequent risk imposed on a prey item by the close proximity of another prey. Arrows indicate track of crow; triangles indicate preys; and circular dots the place from which a prey is seen. The radius of the stippled area represents the Direct Detection Distance, variations in which are indicated by the vagueness of this area's boundary. The area-restricted searching of the predator consequent upon its finding the first prey increases the risk for any subsequent prey in the hatched area, the radius of which equals the distance that could be covered by the searching predator before it gives up (the 'Effective Detection Distance'). Owing to variations in the actual path followed, this distance is likewise variable, but will almost invariably exceed the Direct Detection Distance.

to what extent the search was prey-specific as well as area-restricted.

2. When we next consider whether, with natural populations, this pressure ever comes into play, we are on less certain ground. Asking this question amounts to asking whether a density larger than the one observed (as would occur through a failure of the spacing-out mechanisms) would result in higher predation before other conceivable penalties (such as starvation, or absence of free breeding sites) would occur; and if so, if this higher predation would be due to the predator finding more prey items within reach of its 'Effective Detection Distance'.

Mechanisms promoting the spacing-out of individuals or breeds even within a suitable habitat occur of course widely in nature. Even at the present unsatisfactory state of knowledge we can say that this penalty varies from species to species (see, for instance, Hinde, 3). For instance in some songbirds the first penalty of a larger than normal density may well be starvation; whereas in some seabirds the first penalty may be loss of nest site rather than starvation. Even within a species the critical penalty may well vary with the circumstances, and it is therefore perfectly possible that the predator pressure demonstrated in our experiment would never or rarely be the critical factor in natural populations. This is where the elimination argument applies: we believe that none of the critical factors as yet conceived are likely to be the main ultimate factors forcing well-camouflaged animals to live as far apart as they do, and that predator pressure of the kind proposed may well be the critical penalty.

3. Since living well spaced-out must be a help in evading predation in any animal, the question arises whether it is justified to emphasise in particular its importance for camouflaged animals. If no other selection pressures would apply, and other things being equal, less well camouflaged or even conspicuous animals (in this respect species could of course be arranged on a sliding scale) should be under even stronger pressure to space out. However, other pressures do apply, and more conspicuous animals, having as a rule more direct anti-predator defences than cryptic animals, can afford to live more densely and so utilise for instance food supplies more completely. Our study does not attempt to give information on this, but is concerned with the search for an environmental pressure which could account for the discrepancy, which we believe exists in many camouflaged animals, between the short Direct Detection Distance and the larger inter-individual distances observed. The question whether or not a similar discrepancy exists in non-camouflaged animals must be left open, but seems less likely to be relevant.

4. The question discussed so far, viz. whether it is of advantage to

individuals to live well away from others, which is a simple question of survival value, has to be well distinguished from the evolutionary question at which level the selection pressure acts that produces or stabilises the spacing-out mechanisms. This question arises when it is realised that there are on the one hand species in which individuals space themselves out, or in which parents deposit their brood well away from those of other parents of the same species, and, on the other hand, species in which the parents take care to space out the individual members of their own brood. The larvae of the Peppered Moth quoted above are examples of the first category, and so are many territorial ground breeding birds with cryptic broods; the Sphyngid Moths spacing out their individual eggs are examples of the second. Our experiments apply only to the first category because they differed from the natural situation in one important aspect: our sample eggs were so conspicuous that they were detected from a considerable distance, and as our figures show, the C-plots had as good a chance of being discovered as had the S-plots. But in the natural situation there is no such advertising by conspicuous samples, and, the Direct Detection Distance being short, it must take predators longer to detect, in an area of a given size, the first item of a crowded population than that of a spaced-out population of equal size. This advantage to the C-population might well reduce or even nullify the disadvantage shown in our tests. Because we eliminated this aspect in our tests (which we had to do for practical reasons— it would have been very difficult to attract wild Crows to our plots at all if no sample eggs had been provided) we cannot say whether spaced-out populations would, as wholes, suffer less than crowded populations. Thus, while our experiments suggest a possible selection pressure favouring individuals who space out themselves, or parents who deposit their brood well away from other broods, they do not show whether parents who space out the individual members of their brood will be more successful than those who would not do this. In order to decide this, experiments without advertisement by conspicuous samples would have to be done.

5. With experiments of this kind one could in principle determine for a given prey in a given habitat with respect to a given predator which density would be 'critical' from this point of view—that is, at which density the finding of one prey item would not endanger its nearest neighbour. It is probable that even with inter-egg distances of 800 cm the density was still higher than this 'critical density', for in most tests more than one of the S-eggs were found. But even if we would, by a much wider range of densities and many more tests, have established this 'critical density', we would not be entitled to take

this density as a yardstick for conditions in natural populations, since the 'critical density' varies of course with, e.g. the degree of crypsis of the prey, with the nature of the habitat, and with the persistence and efficiency of the predator, which in turn might depend in part on, e.g. the size and on the palatability of the prey.

There are further reasons to expect that natural populations do not live at the critical density as defined above, for density depends on other ultimate factors as well. Some of these may require an even lower density than that at which increased predation through crowding begins—for instance food supplies may be scattered more widely, and the dispersion mechanism of many species are undoubtedly adapted to the distribution of this vital resource. On the other hand, a species may be subject to pressures towards a higher density than would be desirable in the context of our experiments. Thus Kruuk (6) and Patterson (8) have shown that Black-headed Gulls, whose eggs are protected by their camouflage (Tinbergen et al., 16) nevertheless profit from crowding—this makes the attacks of the parents on some predators more effective. The actual colony density is undoubtedly a compromise between at least these two counteracting pressures. It is very likely that similar considerations apply to very many animals, as they certainly do to many aposematic animals, which definitely clump.

Summary

The paper is concerned with the tracing of a selection pressure which would account for the fact (believed to be sufficiently well established) that individuals of many well-camouflaged species live further away from other individuals of their species than the distance from which even bird predators are able to detect them. Artificially camouflaged Hens' eggs were laid out in plots of different densities. Wild Carrion Crows were attracted to each plot by a standard "sample egg' which, while painted in the same way as the other eggs on the uppermost half, was laid out in a more conspicuous way. In spite of the fact that the Crows spent more time searching in the 'scattered' than in the 'crowded' plots, the crowded eggs suffered a much higher mortality. It is concluded that even for individuals of a well-camouflaged species it must be of advantage to live further away from others than the Direct Detection Distance of their predators. However, the experiments do not show that a crowded population as a whole suffers higher predation than a scattered population; experiments to test this and other aspects of the problem are in progress.

It is argued that the absolute values of the density dependent mortality scores of the experiments cannot be applied to natural populations, because their density will in most cases be determined by other ultimate factors as well.

REFERENCES

1 COTT, H. (1940). *Adaptive Coloration in Animals*, London.
2 CROZE, H. J. (1970). 'Searching Image in Carrion Crows', *Zt. f. Tierpsychol. Suppl.*, **5**.
3 HINDE, R. A. (1956). Papers on territory in *Ibis*, **98**.
4 KETTLEWELL, H. B. D. (1955). 'Selection experiments on industrial melanism in the Lepidoptera', *Heredity*, **9**, 323–42.
5 —— (1956). 'Further selection experiments on industrial melanism in the Lepidoptera', *Heredity*, **10**, 287–301.
6 KRUUK, H. (1964). 'Predators and Anti-predator Behaviour of the Black-headed Gull (*Larus ridibundus* L.)', *Behaviour Suppl.*, **11**.
7 LACK, D. (1955). *The Natural Regulation of Animal Numbers*, Oxford.
8 PATTERSON, I. J. (1965). 'Timing and spacing of broods in the Black-headed Gull (*Larus ridibundus* L.)', *Ibis*, **107**, 433–60.
9 RUITER, L. DE (1952). 'Some experiments on the camouflage of stick caterpillars', *Behaviour*, **4**, 222–32.
10 —— (1955). 'Countershading in caterpillars', *Arch. néerl. Zool.*, **11**, 1–57.
11 TINBERGEN, L. (1960). 'The natural control of insects in pinewoods', *Arch. néerl. Zool.*, **13**, 259–379.
12 TINBERGEN, N. (1953). *Social Behaviour in Animals*, London.
13 —— (1956). 'On the functions of territory in Gulls', *Ibis*, **98**, 401–11.
14 —— (1957). 'The functions of territory', *Bird Study*, **4**, 14–27.
15 —— (1965). 'Behaviour and Natural Selection', in J. A. MOORE (ed): *Ideas in Modern Biology*, New York, 521–42.
16 —— G. J. BROEKHUYSEN, F. FEEKES, J. C. W. HOUGHTON, K. KRUUK and E. SZULC (1962). 'Egg Shell Removal by the Black-headed Gull *Larus ridibundus* L.', *Behaviour*, **19**, 74–118.

343